Molecular Dynamics for Materials Modeling

The book focuses on the correlation of mechanical behavior with structural evaluation and the underlying mechanisms through molecular dynamics (MD) techniques using the Large-scale Atomic/Molecular Massively Parallel Simulator (LAMMPS) platform. It provides representative examples of deformation behavior studies carried out using MD simulations through the LAMMPS platform, which provide contributory research findings toward the field of material technology. It also gives a general idea about the architecture of the coding used in LAMMPS and basic information about the syntax.

FEATURES:

- Provides a fundamental understanding of molecular dynamics simulation through LAMMPS
- Includes training on how to write LAMMPS input file scripts
- Discusses basics of molecular dynamics and fundamentals of nanoscale deformation behavior
- Explores molecular statics and Monte Carlo simulation techniques
- Reviews key syntax implemented during simulation runs in LAMMPS, along with their functions

This book is focused on researchers and graduate students in materials science, metallurgy, and mechanical engineering.

Molecular Dynamics for Materials Modeling

A Practical Approach Using LAMMPS Platform

Snehanshu Pal and K. Vijay Reddy

CRC Press
Taylor & Francis Group
Boca Raton London New York

CRC Press is an imprint of the
Taylor & Francis Group, an **informa** business

Designed cover image: © Shutterstock Images

First edition published 2024
by CRC Press
2385 NW Executive Center Drive, Suite 320, Boca Raton FL 33431

and by CRC Press
4 Park Square, Milton Park, Abingdon, Oxon, OX14 4RN

CRC Press is an imprint of Taylor & Francis Group, LLC

© 2024 Snehanshu Pal and K. Vijay Reddy

ISBN: 978-1-032-34719-6 (hbk)
ISBN: 978-1-032-34720-2 (pbk)
ISBN: 978-1-003-32349-5 (ebk)

DOI: 10.1201/9781003323495

Typeset in Times LT Std
by Apex CoVantage, LLC

Contents

About the Authors

Snehanshu Pal is presently working as an associate professor in the Department of Metallurgy and Materials Engineering, Indian Institute of Engineering Science and Technology, Shibpur, West Bengal, India. He previously worked at the National Institute of Technology (NIT), Rourkela, India for nine years (2014-2023). He has served as a postdoctoral fellow in the Department of Materials Science and Engineering, the Pennsylvania State University. He received his PhD in metallurgical and materials engineering from the Indian Institute of Technology, Kharagpur, India, in 2013. A passionate researcher, critical thinker, and committed academician, Snehanshu Pal currently holds an assistant professor position at the Metallurgical and Materials Engineering Department of NIT, Rourkela, since 2014. His research focuses on the study of the deformation behavior of nanostructured material using MD simulation and modeling of metallurgical processes. He is eager to teach and pass on knowledge and is a highly motivated, reliable, dedicated, innovative, and student-oriented teacher in the fields of mechanical metallurgy, metallurgical thermodynamics, and atomistic modeling of materials.

Snehanshu Pal leads the Computational Materials Engineering and Process Modeling Research Group at NIT, Rourkela, a group dedicated to realizing the underlying physics behind the mechanical behavior of materials and simulating metallurgical processes (http://www.snehanshuresearchlab.org). He has published more than 100 high-impact research articles in internationally reputed journals. He has supervised three doctoral theses and several master's theses. He is an investigator of numerous sponsored research projects and industrial projects. He has active research collaborations with esteemed universities across the globe (such as the University of Florida, the University of Manitoba, Université Lille, and the National Academy of Science of Belarus). In addition, Snehanshu Pal is associated with various esteemed technical and scientific societies such as the Indian Institute of Metals and Indian Institute of Engineers.

K. Vijay Reddy is a postdoctoral researcher in KU Leuven, Belgium, working primarily in the field of computational materials engineering. He did his PhD at the Department of Metallurgical and Materials Engineering, National Institute of Technology (NIT), Rourkela, India, working on nanoscale behavior of materials and design of nano-processing techniques using atomistic simulation techniques. His doctoral research focuses on investigating the material processing of nanoscale metallic systems using molecular dynamics simulation processes. Apart from the doctoral research field, he has worked on multiple research projects and published more than 30 research articles in high-quality journals over the years. He has demonstrated a strong command of computational skills, has been involved in developing many in-house simulation codes, and has gathered vast knowledge from all of his research experiences. Together with Dr. Snehanshu Pal, he has also been associated with various collaborations with esteemed universities across the globe (such as the University of Florida, University of Manitoba, and University of California Irvine).

Apart from atomistic simulations, he has also worked with industrial collaborator Dr. Chandan Halder (manager, Mishra Dhatu Nigam Limited) in the field of microstructure modeling. Vijay Reddy is an integral part of the Computational Materials Engineering and Process Modeling Research Group that is led by Dr. Snehanshu Pal at the National Institute of Technology, Rourkela, a group dedicated to realizing the underlying physics behind the mechanical behavior and processing of materials and simulating metallurgical processes (http://www.snehanshuresearchlab.org).

Preface

Atomistic modeling of materials is a rapidly progressing research field, as it has emerged as a potential tool to assist experimental and theoretical investigations. Moreover, advancement and technological development of computational devices and methods have enabled flexible and easy analysis of larger systems at the atomic scale. Although the foundation of computational methods is based on empirical relations and predictions, the outputs have been always reliable and have benefitted numerous disciplines of science. In the field of atomic-level modeling, molecular dynamics (MD) simulation is a mainstream, powerful, and dependable modeling technique used to understand intrinsic material behavior at the nanoscale level. For example, complex deformation mechanisms in extreme conditions such as high temperature and pressure can be analyzed safely by means of a computer. Meanwhile, nanostructured (NS) materials are recognized as a novel class of materials, which have a strong potential in designing and developing materials with higher strength and ductility and other new application-specific materials with unusual customized properties. These materials, being polycrystalline in nature and with a grain size in the nanometer range, can provide a massive opportunity for scientific applications due to their non-conventional properties. This generates massive interest in exploring the basic structure–property relationships and deformation behavior of polycrystalline single or multiphase homogeneous and heterogeneous materials. From this perspective, MD simulation is a significantly advantageous technique to gain insight and a detailed understanding of the mechanical behavior of NS materials and associated underlying deformation mechanisms at the atomic scale. It has a strong capacity to provide knowledge and information so that blueprinting of experiments becomes easy as well as effective, and the actual number of experiments could be reduced significantly. It is even possible to determine very intricate details of structural characterizations and their effect on the mechanical properties of NS materials through MD simulation, but for some cases, such a degree of dynamic detailing cannot be attained through real experiments. For instance, the evaluation of the role of grain size; grain boundary; triple junctions; and defects such as stacking faults, dislocations, and dislocation loops in performance during deformation of NS materials and associated rationality at the atomic level can be achieved by MD simulations. At this point, we realized the necessity of a book that focuses on a detailed comprehension of the correlation of mechanical behavior with structural evaluation and the underlying mechanism through molecular dynamics techniques using the Large-scale Atomic/Molecular Massively Parallel Simulator (LAMMPS) platform. LAMMPS is a classical MD simulation code that emphasizes material modeling, specifically understanding the material behavior of metals, semiconductors, polymers, and biomolecules. Through this book, our prime purpose is to draw the attention of the readers and make possible the seamless transfer of knowledge by providing representative examples of deformation behavior studies carried out using MD simulations through the LAMMPS platform, which provide research findings that contribute to progress in the field of NS material technology. Therefore, the first chapter is dedicated to a basic theoretical understanding of molecular dynamics simulation and the architecture of code for the LAMMPS platform. The second chapter elaborates upon the types of evaluation that can be performed using MD simulations. After that, Chapters 3 to 6 explain the nanoscale deformation behavior of various metallic systems, glasses, and composites and the enhancement of mechanical properties through grain boundary engineering through the LAMMPS platform. Finally, Chapter 7 deals with the study of material processing, specifically the nanoscale rolling deformation behavior of metallic systems. As a whole, a serious attempt is made to coherently present research efforts that are exceedingly representative of the impact of atomistic simulation-based studies on the understanding of change in structural features during the deformation process and the underlying deformation mechanism of NS materials. This book aims to provide a platform to researchers and technologists interested in developing NS materials and studying their mechanical behavior.

Acknowledgments

The authors would like to take this opportunity to extend their heartfelt gratitude to the National Institute of Technology (NIT), Rourkela, India, and its bright and beautiful-minded pupils. Foremost, we would like to express our genuine thanks to our students Mr. Pokula Narendra Babu, Ms. Srishti Mishra, and Ms. Mouparna Manna for their support and participation. We would like to acknowledge everybody in the Metallurgical and Materials Engineering Department of NIT, Rourkela, India. We would also like to thank the generous support and suggestions from different government and private organizations. We are indeed obligated to the entire budding and beautiful Computational Materials Engineering and Process Modelling Group at NIT, Rourkela, India. Last but not least, we would like to show gratitude to everyone on the publishing team.

1 Atomistic Simulation
A Theoretical Understanding

1.1 INTRODUCTION

One of the great theoretical physicists and scholars, Richard Feynman, said, "I, a universe of atoms, an atom in the universe", which aptly explains that everything in this universe is alike and acts in related behavior at the core level. However, individual versatility and adaptability due to personal emotions, philosophies, and ideologies give every one of us a unique character. Fortunately, an atom and its constituents do not sway by these variables and always follow a fixed set of principles, that is, the classical and the quantum rules of nature. Matter constituted of atoms can be affected only by four fundamental forces of the universe: electromagnetic, strong nuclear, weak nuclear, and gravitation [Lee, 2016]. To understand materials, the only interaction force that needs to be considered is electromagnetic forces, while the other three interactions can be neglected. Atomistic simulation of materials is the study of these electromagnetic interactions between electrons, nuclei, and atoms using computational techniques and methods. Specifically, designing and simulating a material requires the following steps to be followed:

- First and foremost, we need to understand what we want to calculate and analyze. We can define our model and design the material system based on the decision. For instance, a solid material model is designed and analyzed if the requirement is to study the structural properties.
- After deciding what to study, the next step is to fix a simulation platform where a program can be written and executed. The simulation package is selected based on the length and time scale, type of system, and electromagnetic forces. There are numerous platforms, such as density functional theory (DFT) platforms, to study properties at an atomic/molecular level and molecular dynamics (MD) and Monte Carlo (MC) for the nanoscale level. Specifically, for most metallic systems at the nanoscale, the Large-scale Atomic/Molecular Massively Parallel Simulator (LAMMPS) platform is used, whereas for organic systems, packages such as GROMACS are used.
- After deciding on the simulation platform, the next step is to create the model accurately using the selected platform such that it represents the actual system.
- Once the atomic model has been created, then define and assign the interatomic interaction forces between the atoms in the model. Physical rules such as classical or quantum mechanics, energy minimization, and equilibration algorithms should also be finalized and implemented.

DOI: 10.1201/9781003323495-1

- Run the simulation, obtain the output data, and analyze the results. If necessary, run multiple simulation programs with refined parameters and conditions to obtain accurate results.
- Compare the obtained simulated results with those of theoretical studies or experimental analysis.

These steps are a general procedure for atomistic simulation of materials. However, simulation techniques are always a continuous compromise between the accuracy of the results and the speed of performing a simulation. Hence, we need to choose between a small model and longer runtime or a larger (more accurate) system with decreased runtime. If one of the variables between speed and accuracy is not compromised, then the overall expense of performing atomistic simulation will increase drastically. From this perspective, the user has to decide the most suitable method to perform the simulation or segregate the entire simulation model into smaller simulation steps to increase efficiency and reduce time. On the other hand, the platforms on which the simulation is to be run should be selected efficiently. For instance, a DFT simulation can handle only a few atoms for the simulation, whereas the number of atoms in an MD simulation can reach up to a few hundred (or even a million). Reversing the number of atoms in DFT and MD can have a drastic impact on the efficiency and cost of the simulation. Apart from the simulation platform, computational resources also play an important role in effectively executing a simulation program. Hence, the user should be wary of the extent of the computational resources available at their disposal. In this book, we will mainly concentrate on MD simulations and related modeling techniques using LAMMPS for a better understanding of how to program a model, run it, and analyze the results. But before that, we need to discuss a few molecular static models that are implemented during an MD simulation to obtain a stable initial structure of the material, which in turn helps in decreasing the total runtime and increasing the efficiency.

1.1.1 MOLECULAR STATICS

Molecular statics is a constrained-optimization technique based on iterative methods to minimize the potential energy of the pre-defined computational cell by re-scaling the positions of the atoms without disturbing the overall structure of the atomic configuration. The technique is implemented on a system that has a periodic boundary condition in all three dimensions, which will be later discussed in detail in this chapter. To perform the molecular statics technique, there are a few methods such as steepest descents, conjugate gradient, and the Newton-Raphson or quasi-Newton method that are implemented to obtain a minimum energy structure of the model. Of these methods, the most-used energy minimization method for metallic systems is the conjugate gradient method, which is already implemented in the LAMMPS platform. It begins by identifying the direction of the local minimum and then converges towards that position. After reaching the local minimum, the method restarts its identification algorithm and searches for the new local minimum based on the new energy gradient and previous identification direction. This method is faster in comparison to the other methods, as it is rather straightforward in identifying the local minimum and comparatively takes fewer steps to reach the final energy configuration.

1.1.2 MONTE CARLO SIMULATION

The Monte Carlo technique is another statistical method that is used in minimizing the energy of a system by altering the position of the atoms. Conceptually, the algorithm is based on statistical mechanics and lets the atoms jump their places randomly and position themselves in new, low-energy positions, which would eventually decrease the total energy of the system. The MC technique uses discrete and random walks for sampling and a probability factor, $\exp(-U/k_BT)$, known as the Boltzmann factor, where U is the energy state, and k_B is the Boltzmann constant. This method is also implemented in the LAMMPS platform as the gcmc-Grand Canonical Monte Carlo command. The command performs MC calculations at a specific temperature (T) and chemical potential defined by the user. Overall, these molecular static techniques are used in re-arranging the atoms in the system by simple calculations to provide a low-energy and stable structure.

1.1.3 MOLECULAR DYNAMICS SIMULATION

Before diving into the concepts of molecular dynamics, we must first discuss the basis on which the MD theory was developed, the famous classical equation of motion introduced by Sir Isaac Newton in 1686:

$$F = ma \qquad (1.1)$$

where F is the force vector, m is the mass of the atoms, and a is the acceleration vector. This simple and eloquent equation can define the motion of any particle or object, for instance, as small as an atom or as large as the sun and the planets. Molecular dynamics is an atomistic simulation method in which the time evolution of the system (consisting of atoms or molecules) is achieved from the integration of Newton's equation of motion [Lee, 2016]. As we are dealing with atoms and molecules, it is important to make it clear that atoms are assumed to be solid and spherical in shape, and the presence of sub-atomic particles such as protons and electrons is neglected. Broadly speaking, MD simulations are implemented through the following steps:

- First, the initial variables: the atomic positions and velocities are calculated using Equation (1.1) and the provided interatomic potential (discussed in detail in Section 1.3).
- After knowing the initial variables, the system is progressed towards a lower energy state using time steps, as a result of which the position and the velocity of the atoms change.
- After attaining new positions, the first step is again repeated until an equilibrium state of the system is achieved. It also means that the properties of the system do not vary with time and remain constant.

Here, we have presented an overview of the steps that have to be followed to work with an MD simulation, which will be elaborated upon in this book as we go through the chapters. MD, as a simulation tool, was first introduced by Alder and Wainwright, where they simulated the equilibrium phase transition behavior of a few hard-sphere systems [1957]. Ever since then, rapid growth has been observed in terms of the development of effective computer algorithms and advancement in computational power to simulate larger complex systems. For instance, Abraham et al. simulated a system containing up to a billion atoms and studied the failure mechanism using the Teraflop ASCI parallel computing system [2002]. Later, Kadau et al. simulated an astonishing 320-billion–atom system on Livermore's BlueGene/L architecture that contained 131 072 IBM PowerPC440 processors [2006]. In recent years, MD simulation has become a reliable and effective method for investigating various atomistic mechanisms and the underlying physics of numerous nanoscale processes [Borovikov et al., 2015; Mianroodi et al., 2016; Yildiz and Kirca, 2018; Pal et al., 2021]. Moreover, with the use of discrete and small-time steps, frame-by-frame analysis of various material processes can be performed to get a deep insight into the respective phenomenon. Along with this, investigations using MD simulation are flexible enough for dynamic studies at the nanoscale level, which can be very challenging in the case of experimental studies. This simulation method also has numerous advantages, such as reproducibility, repeatability, high resolution, and optimum usage of computational resources. Due to this, MD has broad application in the fields of material science, physics, chemistry (for example, biochemistry, electrochemistry, etc.), and engineering. From the perspective of material science, there are several MD-based studies on the deformation behavior of nanomaterials under static and dynamic loading conditions [Luu et al., 2020] and on heat treatment and solidification processes [Trady et al., 2016; Reddy et al., 2017; Scudino and Sopu, 2018], and they have been reported in the literature. These MD simulations are usually carried out on the LAMMPS platform, which is a classical MD code for modeling materials developed at Sandia National Laboratories [Plimpton, 1995]. Even though the MD simulation method is extremely fast, can handle large specimens, and can approximately predict the bulk properties of materials, it has a few limitations that restrict its widespread use in the field of material science, for instance, availability and accuracy of the interatomic potentials, limited length and time scale, and inability to study functional properties such as electromagnetic and dielectric properties. That said, it is well suited and extensively applied to investigate the mechanical, structural, and chemical properties of nanomaterials.

1.2 GENERAL STEPS OF MD SIMULATION

1.2.1 Initialization of Input Parameters

To run an MD simulation, some initial input parameters should be defined to minimize the errors and obtain consistent results. First, the system type and size should be decided, which includes the lattice structure, box dimensions, and initial position and velocity of all the atoms. The initial positions of the atoms for a crystalline system are assigned according to the crystal structure and lattice parameters of the material chosen. On the other hand, the initial velocity of the system is randomly chosen per the Maxwell–Boltzmann distribution and is a function of the temperature [Lee, 2016]:

$$P(v) = \left(\frac{m}{2\pi k_B T} \right)^{\frac{1}{2}} \exp\left(-\frac{mv^2}{2k_B T} \right) \tag{1.2}$$

where k_B is the Boltzmann constant. Moreover, the direction of the velocities of atoms is assigned randomly such that the total linear momentum of the system is zero. Timestep and total runtime are also defined to perform the simulation efficiently. The timestep is the smallest span of time between two simulation runs and is used to capture the trajectory (both position and velocity) of the moving atoms. Usually, the atoms in the crystalline lattice vibrate in the range of 10^{-13} to 10^{-14} s and hence the time step is considered in the range of femtoseconds (10^{-15} s) to accurately predict the atom trajectory. Apart from this, the boundary conditions and the ensembles are also pre-assigned to define the state of the system.

1.2.2 INTEGRATION OF NEWTON'S EQUATION OF MOTION

Once the force acting on the atoms is decided through the interatomic potential, the next step is the integration of Newton's equation of motion to calculate and update the positions and velocity of the atoms in the system. For this, the velocity Verlet algorithm is an extensively used algorithm, where the atomic positions and velocities at time $(t + \Delta t)$ are attained from the given quantities at time t. First, the velocity (v) is advanced by a half-timestep $(\Delta t / 2)$, and the position (r) is advanced for a full timestep using the half step velocity $v\left(t + \dfrac{\Delta t}{2} \right)$, as represented in the following:

$$v\left(t + \frac{\Delta t}{2} \right) = v(t) + \frac{1}{2!} a(t) \Delta t \tag{1.3}$$

$$r(t + \Delta t) = r(t) + v(t)\Delta t + \frac{1}{2!} a(t)\Delta t^2 = r(t) + v\left(t + \frac{\Delta t}{2} \right)\Delta t \tag{1.4}$$

After that, the acceleration (a) is advanced for a full timestep using the interatomic potential relation and the advanced position $r(t + \Delta t)$, as shown:

$$a(t + \Delta t) = -\left(\frac{1}{m} \right) \frac{dU\left[r(t + \Delta t) \right]}{dr} \tag{1.5}$$

After the calculation of a and r, the velocity $v(t + \Delta t)$ is calculated using the half step velocity $v\left(t + \dfrac{\Delta t}{2} \right)$ and full step $a(t + \Delta t)$ as follows:

$$v(t + \Delta t) = v(t) + \frac{a(t) + a(t + \Delta t)}{2}\Delta t = v\left(t + \frac{\Delta t}{2} \right) + \frac{1}{2} a(t + \Delta t)\Delta t \tag{1.6}$$

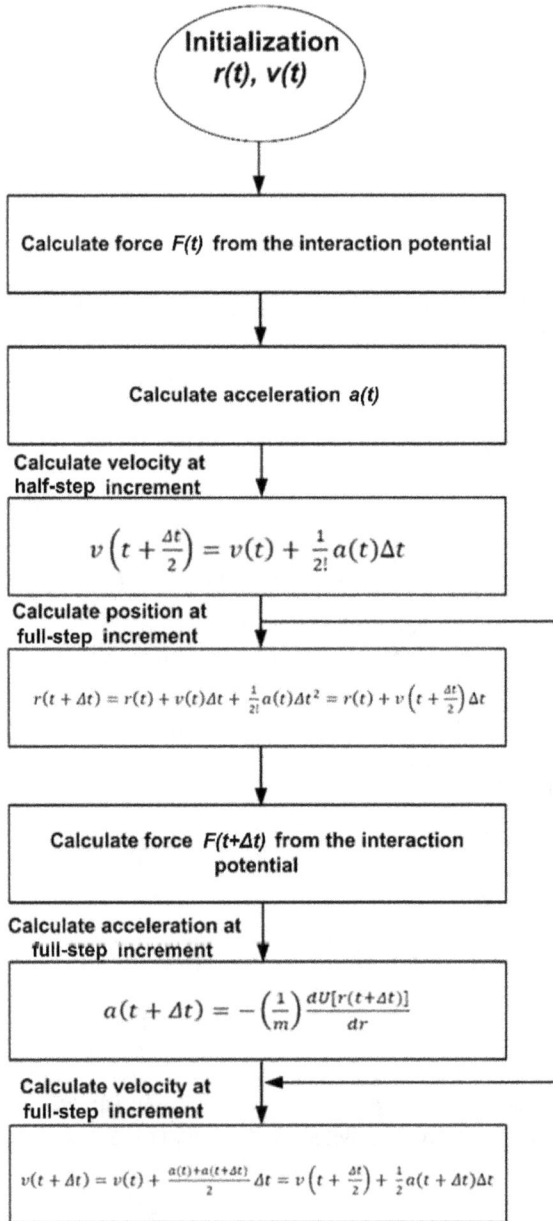

FIGURE 1.1 Flow chart showing the implementation of the velocity Verlet algorithm for a single time step in an MD simulation run.

It is to be noted that the velocities for a full timestep are updated only after updating the positions and accelerations. By implementing such an algorithm, it is ensured that the velocities and the positions are calculated for each and every timestep, and correspondingly the accumulation of error is small. This makes the velocity Verlet algorithm accurate and simple to implement. Moreover, this technique is also time reversible for short and long time periods. A schematic flow chart is presented in Figure 1.1 to illustrate the implementation of the velocity Verlet algorithm during an MD simulation run. We also have to choose and select suitable interatomic potentials, appropriate boundary conditions, and ensembles. These important variables are discussed in detail in the subsequent sections.

1.3 INTERATOMIC POTENTIALS

Interatomic potentials are used to create stability between the attractive and repulsive forces that exist among the nearby atoms in a particular system so that the potential energy of the entire system reaches a minimum value. These potentials are generally formalized through fitting functions to static properties obtained from the experimental studies or the first-principles data [Lee, 2016]. Some of the data obtained from the experimental analysis consists of cohesive energy, modulus values, lattice parameters, the coefficient of thermal expansion, and vibrational data. With the advancement in computational power and efficiency, remarkable progress has been made in the development of various types of interatomic potentials for metals, alloys, semiconductors, ceramics, intermetallic compounds, and organic/polymer materials. Broadly, the interatomic potentials are classified into four categories: pair potential, Tersoff potential, embedded atom method (EAM) potential/modified EAM potential, and ionic solid potential.

1.3.1 PAIR POTENTIAL

Pair potentials are the most simplified form of interatomic potential, which considers the interaction only between two nearby atoms in the system and neglects all the higher-order (three atoms, four atoms, and so on) interactions. The prime example of pair potential is Lennard-Jones (L-J) potential, which is described as [Lennard-Jones, 1925]:

$$U_{LJ}(r) = 4\varepsilon \left[\left(\frac{\sigma}{r} \right)^{12} - \left(\frac{\sigma}{r} \right)^{6} \right] \tag{1.7}$$

where $U_{LJ}(r)$ is the potential energy, ε is the minimum energy of the potential curve, σ is the interatomic distance for a zero potential, and r is the interatomic distance between the atoms. Figure 1.2 shows schematic potential energy vs. interatomic distance plot representing the L-J potential. Such interatomic potentials are used for the simulation of smaller systems such as simple compounds and gaseous systems. For instance, Wei-Zhong et al. investigated the dependence of the self-diffusion coefficient on temperature using L-J potential [2008]. However, for larger systems such as metals and alloys with complex many-body interactions, these interatomic potentials are unsuitable.

1.3.2 TERSOFF POTENTIAL

Tersoff interatomic potential is specifically implemented for simulating covalent solids with strong directional bonds, for example, diamond or silicon. The potential

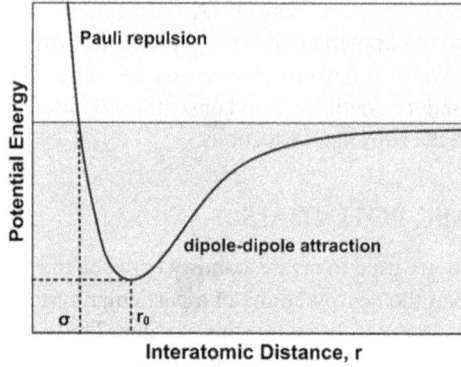

FIGURE 1.2 Schematic representation of the L-J pair potential. r^{-12} term represents the Pauli repulsion, and the r^{-6} term represents the dipole–dipole interaction.

considers the bond angle, coordination number, and bond order in calculating the interatomic forces between atoms along with pairwise calculations. The potential was developed by Tersoff [1988], and it is represented with the following mathematical relation:

$$U_{Tersoff} = \frac{1}{2}\sum_{i \neq j} U_R r_{ij} + \frac{1}{2}\sum_{i \neq j} B_{ij} U_A r_{ij} \tag{1.8}$$

where U_R denotes the repulsive forces, U_A indicates the attractive forces, B_{ij} is the bond order, and r_{ij} is the interatomic distance. Since its development, this potential has been used to investigate systems such as SiC, diamond, Boron carbon nitride (BCN), graphene structure, and many more covalent systems [Chavoshi et al., 2017; Thomas and Lee, 2019; Tranh et al., 2021].

1.3.3 IONIC SOLID POTENTIALS

All the interatomic potentials mentioned previously are typically short range and consider calculating interactions only for a few atoms that come under the cut-off distance. However, the forces in ionic solids can range to very long distances due to polarization and coulombic forces. Hence, the interatomic potential must consider both short-range and long-range interactions. Thus, ionic potentials are developed by considering the potential energy calculation for the nearby atoms and long-range coulombic interactions. Mathematically, this is represented as follows:

$$U_{ionic} = U_{short} r_{ij} + \frac{Z_1 Z_2}{r_{ij}} \tag{1.9}$$

where Z_1 and Z_2 represent the charge values of the two atoms, respectively. On the other hand, U_{short} represents the shortrange interactions, which in turn consist of two additional terms corresponding to the repulsive and long-range attractive forces (Van der Waals force):

$$U_{short}r_{ij} = C_1 \exp\left(-\frac{r_{ij}}{\rho}\right) - \frac{C_2}{r_{ij}^2}$$

(1.10)

where both C_1 and C_2 are constant terms. These potentials are widely used to understand the material behavior of ionic solids such as UO2, saline water, functionalized graphene structures, and ionic liquids [Basak et al., 2003; Koleini et al., 2019; Lodesani et al., 2020].

1.3.4 EMBEDDED ATOM METHOD POTENTIAL

EAM interatomic potentials are widely used for metals and metallic alloy systems, as they take into account both the interaction between pair potentials and the embedded energy due to the N-atom system. In principle, valence electrons of metallic systems are delocalized and constitute the electron density, which impart additional attractive forces to the atom core, which has a positive charge. The embedded energy is the amount of energy required to insert a positively charged core into the electron density. Hence, the EAM interatomic potential takes this additional force into account and calculates the interaction between the atoms in a metal. Mathematically, it is represented with two terms, a pair potential term and an additional embedded energy term:

$$U_{EAM} = \sum_{i<j} U_{ij}r_{ij} + \sum_i F_i \rho_i$$

(1.11)

where $F_i\rho_i$ is the function corresponding to the embedded energy, ρ_i is the electron density, and r_{ij} is the interatomic distance. Figure 1.3 illustrates a schematic representation of an atomic system for EAM potential. Numerous EAM interatomic potentials have been developed for metals (especially cubic crystals like face-centered cubic (FCC) and body-centered cubic (BCC)), transition elements, and a few alloy systems [Mendelev et al., 2008; Wilson and Mendelev, 2016; Marinica et al., 2013]. For instance, a recent study by Yan et al. showed that the solid–liquid interfacial free energy of Al was investigated using EAM interatomic potential [2020]. On the other hand, EAM potential has also been effectively used in understanding the sintering mechanisms in Fe nanoparticles [Luu et al., 2020]. Moreover, EAM interatomic potentials have been extensively implemented for metallic systems for investigating deformation behavior, melting and solidification, grain growth, glass formability, thermal properties, and much more. The selected EAM interatomic potentials have precisely predicted bulk material properties such as lattice constants, phase transitions (e.g., FCC-BCC transitions, FCC-hexagonal close packed (HCP) transitions, and HCP-BCC transitions), mechanical properties such as yield strength, modulus value, ultimate strength values, and melting temperatures and have also previously been implemented in the literature. However, EAM interatomic potentials are not suitable for determining electrochemical (oxidation and corrosion), magnetic, and electric properties.

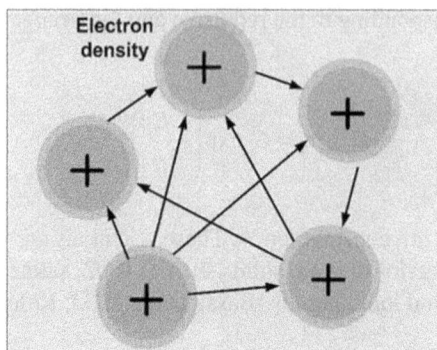

FIGURE 1.3 Schematic representation of a five-atom simulation box showing the embedded positively charged atom cores and the electron density/embedded energy (grey region) to describe the N-atom effect. The arrows indicate the pairwise interactions among the atoms.

1.4 ENSEMBLES

1.4.1 MICROCANONICAL ENSEMBLES

The micro-canonical (NVE) ensemble defines the simulation box as an isolated system and limits the variations of atoms (N), the total volume (V), and the total energy of the system (E). Consequently, the NVE ensemble restricts the exchange of matter and heat in the system with the surroundings. During MD simulation, if the ensemble is fixed to NVE, then the equations of motion considered are Newtonian equations. Moreover, as the total energy is conserved, the trajectory of atoms only results in interchange of the kinetic and potential energy inside the system.

1.4.2 CANONICAL ENSEMBLE

In a canonical (NVT) ensemble, the simulated system is embedded in a heat reservoir with a constant temperature (T) such that thermal equilibrium between the reservoir and the system is maintained. Here, variations in the number of atoms (N), total volume of the system (V), and temperature (T) are restricted. In order to maintain a constant temperature (T), the ensemble is extended with the Nosé–Hoover thermostat [Evans and Holian, 1985]. However, as the system exchanges energy with the reservoir in order to maintain a constant temperature, the total energy of the system is a variable.

1.4.3 ISOTHERMAL–ISOBARIC ENSEMBLE

The isothermal–isobaric (NPT) ensemble restricts the variation of the total number of atoms in the system (N), pressure of the system (P), and temperature (T). In this case, the system is embedded in a heat reservoir to maintain a constant temperature, along with a barostat to maintain a constant pressure. This indicates that there is an exchange of energy (through the reservoir) and volume/work (through the barostat)

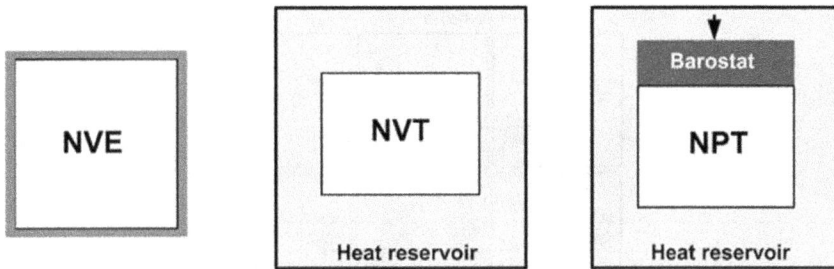

FIGURE 1.4 Schematic illustration of the microcanonical, canonical, and isothermal–isobaric ensembles that are implemented in MD simulation.

from the system. Hence, there can be fluctuations in the total volume and energy of the system while maintaining equilibration. Figure 1.4 shows a schematic diagram of the ensembles that are incorporated in the MD simulation.

After the initialization of the parameter and boundary conditions, the unrelaxed system is equilibrated such that the atoms reach a minimum and constant potential energy value with zero net force on them, or the change in the value of properties (e.g., temperature and pressure) is zero with respect to time. A conjugate gradient algorithm is implemented during the calculations to find the minimum energy configuration of the atoms in the system.

1.5 BOUNDARY CONDITIONS

In MD simulations, the boundary conditions are categorized broadly into periodic boundary conditions and non-periodic/fixed boundary conditions. Depending on the requirements of the simulation model, these boundary conditions are applied accordingly in all three spatial dimensions. For instance, in a rolling process, the transverse direction with respect to the movement of the specimen is considered periodic, whereas the other two directions along the rolling direction are considered non-periodic. However, in the case of a heating and cooling process of a bulk metallic system, all the dimensions are considered periodic. In the next sub-section, detailed descriptions are presented to understand the significance of the boundary conditions in MD simulations.

1.5.1 PERIODIC BOUNDARY CONDITIONS

Periodic boundary conditions (PBCs) are generally implemented to simulate a bulk specimen with a limited-sized simulation box. With this boundary condition, the primary simulation box is replicated in all three directions (x-, y-, and z-axis) as image cells. It is to be noted that the primary simulation box should occupy the space completely during any translational operations. Figure 1.5 shows a schematic representation of the two-dimensional primary simulation box and the image cells.

Image cells

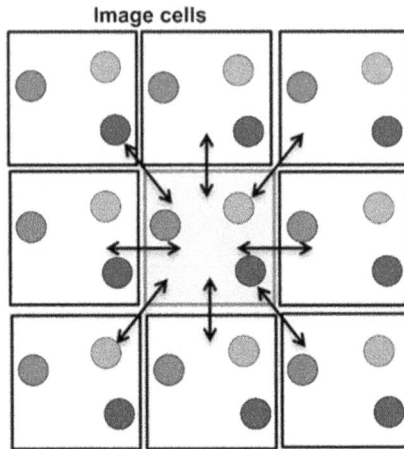

FIGURE 1.5 A schematic representation of the periodic boundary condition. The central cell is the primary simulation box.

Fixed boundary atoms

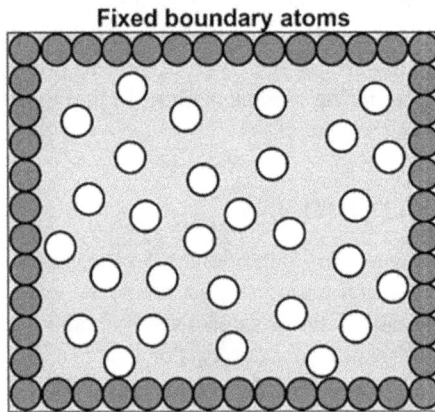

FIGURE 1.6 Schematic illustration of the fixed boundary condition. The gray atoms indicate rigid boundary atoms with zero velocity.

1.5.2 Non-Periodic Boundary Conditions

Non-periodic boundary conditions are categorized into fixed boundary conditions and moving or free boundary conditions. In the case of fixed boundary conditions, the coordinates of the boundary/surface atoms are fixed, making it a rigid body, and the velocities of those atoms are assigned a zero value. Moreover, these boundary atoms are excluded from the MD calculations. Figure 1.6 illustrates a schematic diagram of the fixed boundary conditions using an MD simulation cell.

The free boundary condition is discussed corresponding to its implementation in the LAMMPS platform. In this case, the boundary of the simulation box

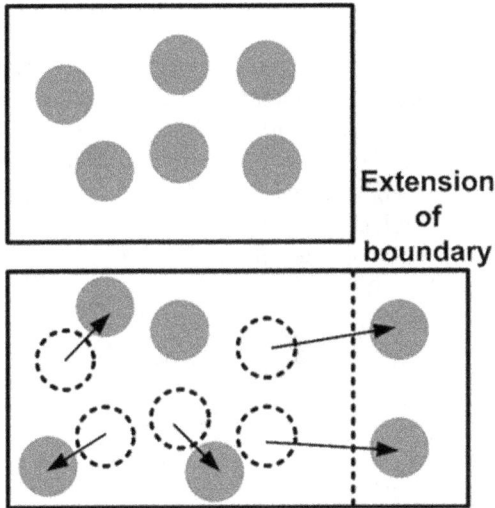

FIGURE 1.7 Schematic illustration of the free/shrink-wrapped boundary condition. The dotted circles represent the previous positions of the atoms (gray solid).

can encompass the moving atom by changing its size along that dimension (i.e., shrink-wrapping). The difference between the fixed and free boundary conditions is that an atom is considered lost for a fixed boundary condition if it crosses the boundary. In comparison, the boundary adjusts its dimensions to include the coordinates of the atom in the case of the free boundary condition. Figure 1.7 depicts a schematic diagram showing the free boundary conditions using a simulation box.

1.6 ARCHITECTURE OF LAMMPS INPUT SCRIPT

In any programming technique, a script consisting of sequence of commands/ instructions is an essential part for smooth operation. Generally, the input commands are read in succession and the operations are executed directly through the compiler or through another program with the help of an executable. In case of the LAMMPS program, the input commands are written in a simple text file (for Windows) and run through an executable file that is available during installation of the package. Now the script can be divided into four sections:

- The first is initialization, where all the initial parameters are defined.
- After initialization, the system definition is the second step, where the dimensions, boundary conditions, and ensembles are defined.
- Once the system is defined, the simulation settings are assigned.
- Last, the output variables are defined and the simulation is run.

We will discuss each of the sections in detail and also describe the important commands that are used in general scripting and executing basic programs.

1.6.1 Initialization

Before designing a model, the first step is to decide on a few basic parameters that will be consistent and define the foundation of the script. A few such relevant commands for LAMMPS input scripts are *units, dimension, atom_style, boundary,* and *echo.* First, the *units* command defines the system of measurements that is used while specifying the dimensions. For instance, the *si* and *cgs* styles for the *units* command assign SI (International System of Units) units and the centimeter–gram–second (CGS) system of units for all variables, respectively. The following is the syntax to write the *units* command:

Syntax	Style
units	*metal*

Here, *metal* is another style in the *units* command, which is specifically used for defining metallic systems and represents parameters such as the mass in terms of gram/mol and energy as electron volts (eV). In total, the units command has eight styles: *lj, real, metal, si, cgs, electron, micro,* and *nano.* Depending on the selection of the style, the timestep for time integration of Newton's equation also varies, so one must be careful in deciding the style of the *units* command based on the model system that is under investigation. Another important command of initialization is *atom_style,* which basically assigns the type of bonding the atoms will have on one another during the simulation. This command allows the programmers to have a variety of atomic bondings and systems to be modeled and shows the wide spectrum of materials that can be investigated in the LAMMPS platform. The syntax for this command is as follows:

Syntax	Style and args
Atom_style	*atomic*

We have listed a few important atom styles and their domain of characteristics/quantities that are stored by each atom during the simulation. These styles are developed internally by Sandia National Laboratories (SANDIA) and are implemented in the LAMMPS platform.

After defining the units and atomic styles, we describe the boundary conditions of the system to be designed. As explained in Section 1.5, there are four different types of boundary conditions that are implemented in a LAMMPS input script. These are: periodic (p), non-periodic and fixed (f), non-periodic with shrink wrap (s), and shrink-wrapped with a minimum value (m). The syntax for this command is as follows:

Syntax	Style
boundary	*p p p*

The boundary conditions are designated to all three axes of the system, and in the previous example, periodic conditions are applied to the x-, y-, and z-axes,

TABLE 1.1

List of Atom Styles and Their Attributes That Are Used to Model a System

angle	bonds and angles	bead-spring polymers with stiffness
atomic	only the default values	coarse-grain liquids, solids, metals
body	mass, inertia moments, quaternion, angular momentum	arbitrary bodies
bond	bonds	bead-spring polymers
charge	charge	atomic system with charges
dielectric	dipole, area, curvature	system with surface polarization
dipole	charge and dipole moment	system with dipolar particles
electron	charge and spin and e-radius	electronic force field
full	molecular + charge	bio-molecules
line	end points, angular velocity	rigid bodies
mesont	mass, radius, length, buckling, connections, tube id	mesoscopic nanotubes
molecular	bonds, angles, dihedrals, impropers	uncharged molecules
peri	mass and volume	mesoscopic peridynamic models
smd	volume, kernel diameter, contact radius, mass	solid and fluid smooth particle hydrodynamics (SPH) particles
sphere	diameter, mass, angular velocity	granular models
spin	magnetic moment	systems with magnetic particles
tri	corner points, angular momentum	for rigid bodies

respectively. We can also choose to apply a combination of boundary conditions on the three axes depending on the requirement of the simulation system. Another command that is generally used during the initialization is the *echo* command, which enables users to output or "echo" the variables on the screen and into a log file so that users can trace changes occurring during the simulation run. This command also prints a message on the screen if the simulation program stops abruptly due to errors in the input script.

1.6.2 Defining the System

In order to work on materials and understand their properties, experimentalists need to prepare specimens of certain shape, size, and volume and then subject them to corresponding treatment to study specific properties. Simulation techniques are no exception to this; after the initialization of the parameters in LAMMPS, our next objective is to describe the material system that is to be studied. In this section of the script, the size of the simulation box, the shape of the specimen, and the crystal orientation of the atoms (if they have an orientation) are described. For this, there are a few commands that are specific for such operations, such as the *lattice, region, create_box,* and *create_atom* commands. For instance, the *lattice* command defines the crystal lattice by describing a set of spatial points (unit cell), which is filled with atoms (using the *create_atom* command) and is repeated in three dimensions to create the entire material. It is to be noted that the *lattice* command can be used only after creating a simulation box space using the *create_box* command. The syntax for the *lattice* command is as follows:

Syntax	Style
lattice	*custom* 3.52 *a1* 1 0 0 *a2* 0.5 1 0 *a3* 0 0 0.5 *basis* 0 0 0 *basis* 0.5 0.5 0.5

Here, using the *custom* style, we can describe and design any type of crystal structure, provided we know the lattice parameter of the crystal and the edge vectors *a1*, *a2*, and *a3* along the x-, y-, and z-axis, respectively. Moreover, we can create multiple crystal structures inside the simulation box using the *region* command, where a specific region in the simulation box is assigned one crystal structure (material) and a different region is assigned another structure. Figure 1.8 shows a schematic diagram where a laminate structure of two different materials is created using the *lattice*, *region*, and *create_atom* commands. Another important command to define the material system is the *read_data* command, which reads in the data file containing information on the atomic coordinates and types of materials. This command enables the user to design the material system separately and then input the data file directly into the script.

1.6.3 SIMULATION SETTINGS

Once the topology of the material is decided, the user will then apply the settings to the simulation box to carry out interactions between the atoms. In order to do so, the

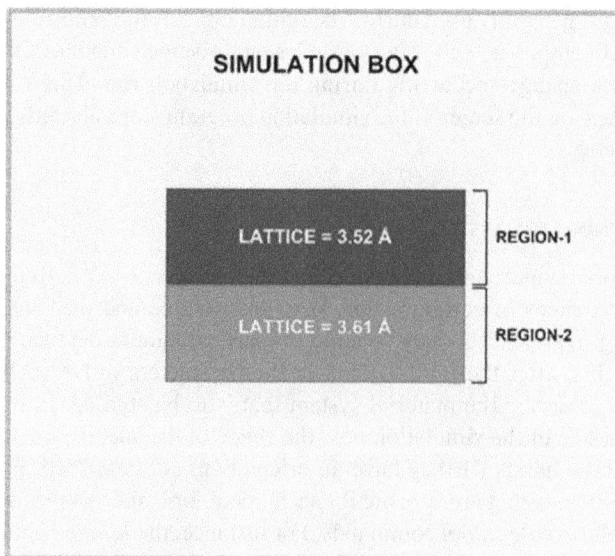

FIGURE 1.8 Schematic illustration showing the simulation box and creation of a nanolaminate specimen using the *region* command. Two different regions have been designated, and crystal structures with varying lattice parameters are used to create a laminated structure.

TABLE 1.2

List of Syntax Used during the Simulation Setting of the Input Script

Syntax	Use	Style
pair_coeff	Defines the pairwise coefficient between atoms (* and * indicate interaction between all atoms).	*pair_coeff* * * Mendelev_Cu2_2012. eam.fs Cu
min_style	This syntax chooses an energy minimization algorithm to perform on the atoms. Here, the example is the conjugate gradient method (cg).	*min_style* cg
run_style	This command chooses the algorithm for the time integration for simulations in LAMMPS.	*run_style* verlet
compute	This command can compute a large variety of properties during the simulation. For instance, the kinetic energy (ke) of all the atoms is calculated. 3 represents an ID for the computation.	*compute* 3 all ke/atom
thermo and *thermo_style*	The *thermo_style* command is used to print thermodynamic data onto the screen and log file during the simulation run. *thermo* determines the frequency of iterations after which the values are printed.	*thermo* 10 *thermo_style* custom step temp vol press pe ke etotal
fix	The *fix* command is used to apply operations on the atoms during the simulation run. For instance, we can assign the type of ensemble that is applied on the atoms using this command. Moreover, this command is also used to implement all types of deformations, thermal fluctuations, shock and impact, and many more operations.	*fix* 1 all nve *fix* cooling all npt temp 100 100 0.01 iso 0.0 0.0 0.1 *fix* mine top setforce 0.0 NULL 0.0
timestep	This command sets the size of the timestep for the MD simulation in LAMMPS.	*timestep* 0.002
dump	This command is used for getting the output file at different frequencies of iterations. Here, 1 refers to the ID of the dump, and 100 indicates that every 100 iterations, a dump/output file (dump. lammpstrj) is created.	*dump* 1 all atom 100 dump.lammpstrj

user has to define the force field coefficients and timesteps and has to perform energy minimization, impose boundary conditions, fix the time integration and diagnostic options, and compute various output parameters that are to be dumped. We have listed the syntax for commands during the simulation settings in the following table.

1.6.4 PROGRAM EXECUTION

After the simulation settings, the final step is to run the simulation for the required number of iterations and time. For this operation, the *run* command is used, and the syntax is as follows:

Syntax	Style
run	*run* 10000 (this is the number of iterations)

1.7 POST-PROCESSING ANALYSIS

Post-processing and visualization of the simulated data obtained from the MD simulation run on the LAMMPS platform can be carried out using a visualization software package, such as OVITO [Stukowski, 2009], VMD [Humphrey et al., 1996], Atomeye [Li, 2003], or ParaView [Ahrens et al., 2005]. Specifically, these visualization tools are used to show the atomic configurations, defect analysis, stress distributions, atomic orientations, atomic movements and displacements, and cluster analysis. Apart from these, these programs are very capable of analyzing a wide range of parameters such as the pressure and temperature distribution and variation in volume during thermal or thermo-mechanical processing. In this section, a detailed description is provided of the post-processing methodology that can be carried out to understand the atomistic behavior of metals. We have restricted our discussion only to metallic systems and explained the methods considering OVITO as the visualization tool, although the other tools are similar to operate.

1.7.1 IDENTIFICATION OF ATOMIC CONFIGURATION

The atomic configurations of the specimens are analyzed in OVITO through common neighbor analysis (CNA) [Honeycutt and Andersen, 1987], centro-symmetry parameter (CSP) [Kelchner et al., 1998], and polyhedral template matching (PTM) [Larsen et al., 2016]. Common neighbor analysis is a tool used for the visualizing the local crystal structure of a specimen. The algorithm used for CNA defines the local atomic structure by using a radial distribution function (RDF) decomposition and categorizes it as crystalline or amorphous. The radial distribution function is mathematically expressed by the following relation [Zhang et al., 2015]:

$$g(r) = \frac{V}{N^2}\left(\sum_{i=1}^{N} \frac{n(r)}{4\pi r^2 \Delta r}\right) \tag{1.12}$$

where V represents the volume, N represents the total number of atoms, and $n(r)$ represents the number of atoms present between a distance of r and $(r + \Delta r)$. CNA is expressed through the decomposition of the previous equation [Faken and Jónsson, 1994]:

$$g(r) = \Sigma_{jkl} g_{jkl}(r) \tag{1.13}$$

On the other hand, the centro-symmetry parameter calculates the local lattice disorder and characterizes a perfect or defective lattice in the specimen. Mathematically, it is represented as:

$$CSP = \sum_{i=1}^{N/2} \left| R_i + R_{i+N/2} \right|^2 \tag{1.14}$$

Apart from these, polyhedral template matching [Larsen et al., 2016] can also be implemented to identify the local crystal structure of metallic specimens (refer to Figure 1.9). The reason behind using the PTM method over CNA is that the former method is better suited to determine the structural evolution during high strain (plastic) deformations [Larsen et al., 2016]. Moreover, apart from the crystal structure, the PTM modifier also calculates the local lattice orientation and stores the values using quaternions, which aids in the orientation analysis of the crystals in the metallic system.

1.7.2 DISLOCATION AND ATOMIC STRAIN ANALYSIS

During any thermo-mechanical or mechanical tests, a material undergoes structural modifications and generates defects such as vacancies, dislocations, and voids/cracks to accommodate the external forces after crossing the threshold limit (elastic limit). However, before crossing any thresholds, the material will gradually develop internal strain but in a non-uniform manner even if the external forces are uniform. Visualization of the atomic strain and the defects (dislocations) along with their spatial distribution gives many insights into the dynamics of the deformation. Specifically, the dislocations are 1-dimensional (1D) defects resulting from mechanical deformation of metals and are crucial in understanding numerous mechanical properties and phenomena. The different types of dislocations that are generated during these mechanical and thermo-mechanical processes are identified and visualized through

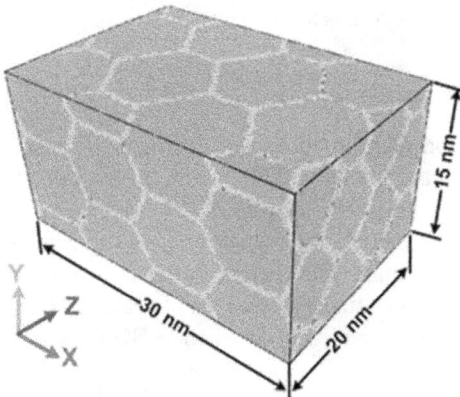

FIGURE 1.9 A nanocrystalline (NC) Ni specimen identified through the PTM method. The atoms are arranged in an FCC crystal structure, and the gray atoms represent grain boundaries with random orientation.

a dislocation extraction algorithm (DXA) [Stukowski et al., 2012], which has already been implemented in the visualization software (e.g., the OVITO package). For instance, various types of dislocations that are generated in an FCC crystal lattice are shown through a line-based representation with different colors such as perfect dislocations being indicated by blue, whereas partial dislocations like Shockley partials are indicated with green. Sessile dislocations such as stair-rod dislocation, Hirth partial dislocations, and Frank partial dislocations are indicated with pink, yellow, and cyan respectively. A few of the dislocations that do not fall under these categories are generally represented with red. It is to be noted that this color classification is specifically for dislocations generated in the FCC lattice. Along with the representations, the DXA also represents the Burgers vector of each dislocation, dislocation type, and total dislocation length in the specimen. Moreover, this algorithm also aids in the separation of a perfect lattice from a defective lattice (with dislocations) by implementing a defect interface mesh. Atomic strain analysis is the calculation of the strain tensor at each of the atomic coordinates from the relative movement of the nearby atoms [Shimizu et al., 2007; Falk and Langer, 1998], which is also implemented through the OVITO platform. Thus, the modifier calculates the deformation gradient with respect to the explicitly quantified initial or deformed (previous iteration values) configurations. In this method, the atomic strain is calculated based on the Green–Lagrangian strain tensor, and mathematically, the shear and volumetric strains are represented as:

$$\varepsilon_{shear} = \left[\varepsilon_{xy}^2 + \varepsilon_{xz}^2 + \varepsilon_{yz}^2 + \frac{1}{6}\left(\left(\varepsilon_{xx} - \varepsilon_{yy}\right)^2 + \left(\varepsilon_{xx} - \varepsilon_{zz}\right)^2 + \left(\varepsilon_{yy} - \varepsilon_{zz}\right)^2\right)\right]^2 \quad (1.15)$$

$$\varepsilon_{volumetric} = \frac{\left(\varepsilon_{xx} + \varepsilon_{yy} + \varepsilon_{zz}\right)}{3} \quad (1.16)$$

where ε_{shear} is the shear strain, and $\varepsilon_{volumetric}$ is the volumetric strain values. The modifier aids in understanding the evolution of shear bands, strain accumulations, and the slip phenomenon in the materials at the nanoscale.

1.7.3 ORIENTATION ANALYSIS

Lattice orientations and grain rotations are identified by implementing an algorithm using polyhedral template matching. The investigation for determining the texture of nano-grained metallic specimens is performed by the post-processing visualization tool OVITO [Stukowski, 2009]. For each atom in the specimen, the orientations are calculated and stored as quaternion values in the form of a matrix. Generally, the quaternion notion that is denoted by q is a set of scalar and vector elements, that is, $[q_0, q_1, q_2, q_3]$, and is mathematically signified as [Krakow et al., 2017]:

$$q = q_0 + iq_1 + jq_2 + kq_3 \quad (1.17)$$

where q_0 indicates the scalar term, and q_1, q_2, q_3 indicate the vector terms. The mathematical expression of Equation (1.17) can also be modified as:

$$q = \langle s, V \rangle \quad (1.18)$$

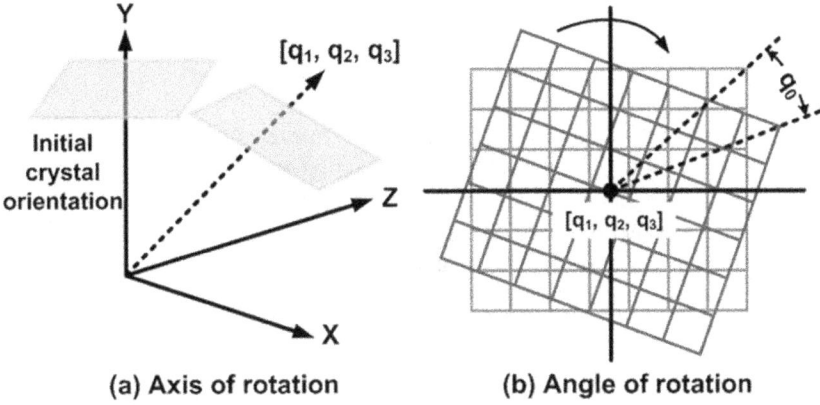

(a) Axis of rotation   (b) Angle of rotation

FIGURE 1.10   Schematic representation of the axis of rotation and the angle of rotation that are computed by implementing the PTM method.

In general, compression and shear loads cause reorientation of the grains, for which the $[q_0, q_1, q_2, q_3]$ values can be linked with the alteration in the rotational axis and the rotational angle through a unit vector ξ by following the mathematical equation [Krakow et al., 2017]:

$$q = \cos\left(\frac{\omega}{2}\right) + \xi_1 \sin\left(\frac{\omega}{2}\right) + \xi_2 \sin\left(\frac{\omega}{2}\right) + \xi_3 \sin\left(\frac{\omega}{2}\right) \tag{1.19}$$

$$\text{or } q = \left\langle \cos\left(\frac{\omega}{2}\right), \xi_i \sin\left(\frac{\omega}{2}\right) \right\rangle \tag{1.20}$$

where rotational angle (ω) is indicated by the scalar term $\cos\left(\frac{\omega}{2}\right)$, and the rotational axis is indicated through the vector term $\xi_i \sin\left(\frac{\omega}{2}\right)$. Figure 1.10 illustrates a schematic representation of the axis of rotation and angle of rotation that are calculated using these equations.

1.7.4   VIRTUAL DIFFRACTION ANALYSIS

The algorithm is implemented in the LAMMPS platform, which generates virtual diffraction patterns from the atomic positions in the specimen [Coleman et al., 2014]. Through this algorithm, a 3-dimensional mesh of reciprocal lattice points is generated on a rectilinear grid. The intensity of the diffracted beam ($I_e(K)$) is then calculated for each of the reciprocal lattice points by evaluating the structure factor ($F(K)$), its complex conjugate ($F^*(K)$), and the Lorentz polarization factor ($L_p(\theta)$). The mathematical representation is as follows:

$$F(K) = \sum_{j=1}^{N} f_j(\theta) \exp(2\pi i K.r_j) \tag{1.21}$$

$$L_p(\theta) = \frac{1 + cos^2(2\theta)}{\cos(\theta) sin^2(\theta)} \tag{1.22}$$

where f_j indicates the atomic scattering factor, r_j represents the atomic positions, and θ represents the angle of diffraction. This simulation technique has been extensively implemented in the characterization of materials through MD simulation [Coleman et al., 2013; Herron et al., 2018]. This algorithm has been implemented to investigate phase transformation, strain accretion, lattice distortions, and amorphization processes by analyzing the atomic diffraction patterns.

1.7.5 Voronoi Analysis

Voronoi polyhedral analysis calculates the Voronoi tessellation of the specimen by taking the atom coordinates as the center of the Voronoi cell and is implemented in the OVITO package [Stukowski, 2009]. Prominently, this method is used to determine the population of various clusters (such as icosahedral and distorted icosahedral) in amorphous solids. Here, the Voronoi polyhedron is represented by indexing and counting the number of edges and faces present in the cell and is described in the form of $<n_3, n_4, n_5, n_6>$, where n represent the number of faces, and the subscript denotes the number of edges. For instance, an icosahedral cluster is a polygon with 12 faces, and each face has 5 edges. Hence, the Voronoi index vector is represented as $<0\ 0\ 12\ 0>$, which is a characteristic signature of an icosahedral cluster. Since the method counts the number of atoms around the central atom (center of the Voronoi cell), this method also aids in the computation of the coordination number for each atom. Moreover, Voronoi analysis has been commonly used in investigating structural evolution in amorphous solids such as metallic glasses [Tang and Wong, 2015; Foroughi et al., 2016; Reddy and Pal, 2017].

2 Physical Property Evaluation by MD Simulation

Understanding the physical properties of nanoscale metals requires deep insight into the relevant scientific principles. Topics to cover include synthesis methods: methods for synthesizing nanoscale metals that can impact their physical properties, structure, and morphology. Nanoscale metals can exhibit a variety of structures and morphologies, such as nanoparticles, nanowires, and nanotubes. These structures can influence the physical properties of the material, such as its melting point, conductivity, and strength. The mechanical properties of nanoscale metals are often different from those of bulk metals due to factors such as surface area, grain size, and defects. Some key mechanical properties to consider include hardness, elasticity, and ductility, Nanoscale metals can have unique thermal properties, such as a high surface area-to-volume ratio, which can impact their melting and boiling points, as well as their thermal conductivity.

2.1 PREPARATION OF NANOSCALE SAMPLES

Metallic nanoparticles can be prepared using various methods, each with its own advantages and intended purposes. Three widely used methods are the sol-gel method, hydrosol/magnetic fluid method, and vacuum deposition method. The sol-gel method involves mixing a metal salt solution, such as $AgNO_3$, with tetraethylorthosilicate (TEOS), ethanol, water, and a catalyst such as HNO_3. The resulting mixture is dispersed and dried and then reduced at a temperature of 400°C in hydrogen gas. This method yields high-purity, isotropic nanoparticles with a low-temperature annealing requirement. It also allows for the removal of water absorbed in the porous gel and hydroxyl groups that affect optical absorption. The hydrosol/magnetic fluid method involves using a reducing agent to embed metal particles in a protective gelatin, resulting in relatively narrow size distribution with an average diameter of 20 Å. The magnetic fluid can also be prepared using Fe_3O4 particles surrounded by oleic acid as a surfactant for protection from aggregation and dispersion in water. The vacuum deposition method involves reducing the momentum of evaporated metallic atoms or clusters by collision with inert gas in a vacuum chamber and lowering the substrate temperature to liquid nitrogen temperature during thermal evaporation. This results in condensed metal atoms that lose van der Waals attraction between particles, preventing further aggregation. The resulting smoke can be collected from the substrate or walls of the evaporation chamber. Another method for producing nanoparticles is ball milling, which is particularly useful for hard and

DOI: 10.1201/9781003323495-2

brittle ceramic materials. This process involves milling powders of around 500 nm in size with tungsten-carbide (WC) spheres, resulting in nanocrystals, non-crystals, and pseudo-crystals that are several nanometers in size. However, one downside of ball milling is the potential for surface contamination and nonuniformity in structure. Nonetheless, it is a relatively simple method for nanoparticle production. To address some of the challenges associated with ball milling, an addition of 1–2% of methanol or phenol can be used to prevent diffusion and the solid reaction of the nanoparticles. Despite its limitations, ball milling remains a popular method for producing nanoparticles due to its effectiveness for hard and brittle materials and its simplicity. Overall, the choice of method depends on the desired properties and applications of the resulting nanoparticles.

With respect to the preparation of nanowires and nanotubes, carbon nanotubes (CNTs) are gaining immense attention as promising material due to their exceptional properties such as rigidity, strength, elasticity, electric conductivity, and field emission. This chapter focuses on growing multi-wall carbon nanotubes on metallic substrates for field emission applications, particularly in display and light illumination. Chemical vapor deposition (CVD) on silicon wafers with a thin film of transition metal catalysts is the most commonly used method for growing CNTs, which produces hexagonally packed graphite on the cooling surface side. However, different production methods result in different genetic characteristics, such as single or multiple walls and conducting or semi-conducting properties, which are crucial factors for the industrial applications of CNTs. Yet the growth mechanism of CNTs remains unclear, including adjusting their size, growth density, and structural qualities. Previous research on CVD-based growth mechanisms of CNTs has emphasized the critical size effect of catalyst nanoparticles on the quantity and quality of CNT production. The size of catalyst nanoparticles deposited on substrates determines whether single-wall or multi-wall nanotubes are produced. Iron group metals' nanoparticles are known to dissolve more carbon elements and reduce the melting temperature, leading to the precipitation of supersaturated carbons on the cooled contact side with the help of curvature-induced surface tension. Researchers have reported a maximum catalyst size of 100 nm for the successful growth of single-wall CNTs. Other methods for CNT production, such as arc discharge and laser ablation, have their advantages and disadvantages. The combination of the catalyst and CVD methods has become the primary approach for growing CNTs, with the catalyst method enhancing growth and the CVD method improving upright alignment and simplifying the products. The radio-frequency hot-filament chemical vapor deposition (RFHFCVD) system involves dissociating reactant gases with a microwave field before entering the chamber, allowing CNTs to grow densely from catalyst particles of about 1 μm. By replacing the deposition catalyst film on the silicon surface with polished Cu-Ni and Cu-Ni-Fe-Co bulk alloys and using RF-induced self-bias, researchers can grow well-aligned carbon nanotubes that potentially serve as electron emission tips with easy manipulation.

Furthermore, bunched and multi-circularly wrapped CNTs are observed to grow on iron group and metal alloys using the microwave-enhanced hot-filament method with a dilute gas of ammonia at a radiofrequency (RF)-induced bias of –200 V. The growth of CNTs on copper-based iron group alloys provides a higher resistance to erosion during high-current emission, with the benefit of larger heat conduction than conventional substrates. Researchers arc melted a copper-based iron group alloy (CuMx, M = Fe, Co, Ni, x = 10–20 wt%) in a vacuum furnace, polished it, and wire cut it into small pieces of size $10 \times 5 \times 0.5$ mm^3. The catalyst iron performs melting, dissolving, and precipitating carbon to elongate the tube and then sucking inside the tube to achieve vapor–liquid–solid (VLS) growth. The VLS model growth mechanism involves hydrogen etching on the surface, formation and rearrangement of nano-catalyst particles on the alloy substrate surface, dissolution of carbon atoms by nano-catalyst particles, and precipitation and lifting up of carbon atoms.

2.2 PHYSICAL PROPERTIES IN NANOSCALE METALS

When a material is reduced to the nanoscale, a significant change in its properties is observed, primarily due to the increase in surface-to-volume ratio. As the size of the material decreases, the surface area to volume ratio increases rapidly. This can be understood by considering a spherical particle with a radius r. The surface area of this particle is given by $A = 4\pi r^2$ and its volume by $V = (4/3)\pi r^3$. The ratio of the surface area to the volume of the spherical particle is inversely proportional to its radius. Thus, as the size of the particle decreases, the surface area–to-volume ratio increases. This concept is depicted in Figure 2.1. The surface area–to-volume ratio plays a crucial role in determining the surface activity of a material. When the surface area of a material increases, it provides more catalytic sites, thereby enhancing its catalytic activity. However, the increase in surface area–to-volume ratio also has an impact on the physical and chemical properties of the material, which are elaborated upon in this chapter.

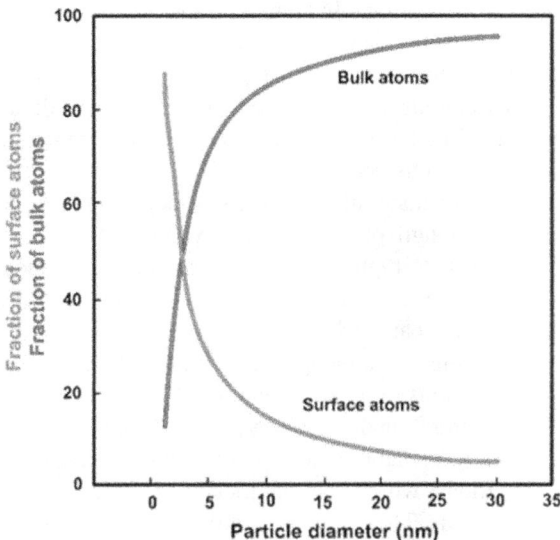

FIGURE 2.1 Atomic fraction vs. grain size plot showing increased surface atom/bulk atom ratio with decrease in grain size.

When a solid substance is melted, it acquires a spherical shape to obtain the minimum energy state. Ionic compounds have higher melting points than covalent bond compounds because more energy is required to break ionic bonds. On the other hand, nanomaterials have lower melting points than bulk materials because of the increased surface energy and reactivity, leading to surface oxidation and a change in surface composition. For instance, the decrease in melting temperature and phase transition temperature of nanomaterials is mainly due to their increased surface area to volume ratio, which results in higher surface energy. The surface of the particles is more reactive and can easily react with the surrounding atmosphere, leading to surface oxidation and a change in surface composition. As a result, the surface atoms or molecules require less energy to break away from the surface and melt, which leads to a lower melting temperature. The change in melting temperature and phase transition temperature has significant implications for the properties and behavior of nanomaterials. For example, the lower melting temperature of gold nanoparticles makes them useful in nanomaterial applications where high temperatures cannot be used, such as in biomedical imaging and therapy. The decreased Curie temperature of ferroelectric nanoparticles may also be useful for designing new types of electronic devices. For example, one-dimensional nanostructures, such as Ge nanowires, also have lower melting temperatures than their bulk forms. In addition to melting temperature, the phase transition temperature also decreases with decreasing size, as demonstrated by the Curie temperature of the ferroelectric to paraelectric transition of barium titanate. In addition, the decrease in melting temperature and phase transition temperature can also affect the stability and durability of nanomaterials. The reduced temperature required for melting may lead to thermal instability and degradation of the nanoparticles during processing or use. Therefore, it is important to consider the size and surface properties of nanomaterials when designing and using them in various applications. In summary, the melting temperatures of nanomaterials are lower than those of bulk materials due to the increased surface energy and reactivity of nanoparticles. This has important implications for the properties and behavior of nanomaterials, as well as for their stability and durability in various applications. Understanding these effects is essential for designing and using nanomaterials effectively in various fields.

The elastic modulus of a material is another size-dependent physical property that is closely related to the strength of the bonds between its atoms or molecules. Higher bond strength corresponds to higher melting points and elastic moduli. The elastic modulus is proportional to the second differential of the interatomic force–distance curve at the equilibrium separation distance. The elastic properties of crystalline materials are generally considered independent of microstructure. However, increasing temperature causes atoms to separate, leading to a decrease in the elastic modulus. Increasing the concentration of defects, such as vacancies, can also decrease the elastic modulus, but only at higher defect concentrations. Nanomaterials have very high defect concentrations, which can significantly reduce their elastic properties compared to bulk materials. The elastic constants of nanocrystalline materials were first measured using samples prepared by the inert gas condensation method. These studies found that the elastic modulus of nanocrystalline compacts was 30%–50% lower than that of bulk materials. One proposed explanation for this is that the elastic moduli of grain boundaries are much smaller than those of the grain bulk regions due to weaker bonds and lower atom density. According to the linear theory of elasticity for polycrystalline materials, the elastic constants are expected to decrease proportionally with the volume fraction of the grain boundaries. However, unlike nano grained materials, the elastic modulus of carbon nanotubes has been found to increase as the tube diameter decreases. This increase in apparent elastic modulus is believed to be due to surface tension effects.

Nanocrystalline materials possess a higher surface area per unit volume fraction of grain boundaries as compared to their microcrystalline counterparts. However, the high-energy nature of grain boundaries results in a driving force for reducing their surface area per unit volume, which can only be achieved through grain coarsening involving the migration of grain boundaries. This process is primarily diffusion limited and dependent on various factors such as temperature and composition. The exponential dependence of grain growth on temperature has been well established for microcrystalline solids and has also been found to apply to nanocrystalline materials. Nevertheless, the activation energy for grain growth in nanocrystalline solids is expected to be lower, and the exponent factor, which is typically 2 for microcrystalline grains, is normally much higher for nanocrystalline grains. These factors result in enhanced grain coarsening kinetics in nanocrystalline materials, making them unsuitable for high-temperature applications. However, researchers have made progress in grain boundary engineering to reduce the grain coarsening kinetics of nanocrystalline materials. One method involves pinning the grain boundaries with secondary particles such as ceramics, which are insoluble in the matrix even at elevated temperatures, to inhibit grain growth. Zener pinning is an effective technique where the particles should be stable at high temperatures and not undergo coarsening themselves.

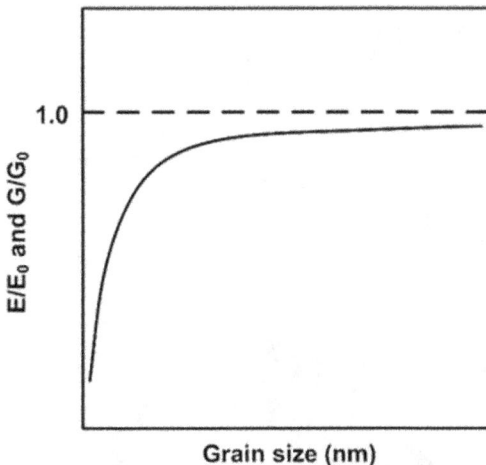

FIGURE 2.2 Schematic diagram showing the change in the modulus value with alteration in the grain size.

Another technique involves enabling a grain boundary with a different composition from that of the bulk material. This approach results in a solute drag effect, where grain boundary migration involves the diffusion of all involved chemical species, acting as a rate-limiting step for grain boundary migration. Heterogeneous composition can be obtained through grain boundary segregation effects, as demonstrated in the case of Cu alloyed with Ag to reduce grain growth tendencies in nano-silver. In conclusion, the service temperature applicability of nanomaterials can be enhanced through effective grain boundary engineering techniques, such as Zener pinning and the solute drag effect, which inhibits grain growth and allows for the utilization of the advantageous properties of nanomaterials.

2.3   EVALUATION OF MECHANICAL PROPERTIES

When it comes to nano-metals, the size of the grains has a big impact on their mechanical strength. The smaller the grain size, the higher the yield stress. The Hall–Petch equation is an important tool for understanding how microstructure affects a metal's mechanical properties. According to this equation, at low temperatures, the yield strength is linearly related to the inverse square root of the average grain size $(1/(d)^{1/2})$. This means that reducing the grain size to the nanoscale can lead to a significant increase in mechanical strength. This has already been demonstrated for various metals and alloys, as shown in Figure 2.3. In fact, nano-metals often exhibit tensile strengths that are two or three times greater than conventional microcrystalline alloys. The strength of nano-metals is not solely dependent on their grain size, as there is a limit to how much the strength can be improved by reducing the grain size. The limit is influenced by various factors such as the chemical composition, melting temperature, and the presence of second-phase particles. This is due to the accumulation, annihilation, and rearrangement of defects generated during processing, as well as the thermally activated phenomena of recovery and grain growth. The effectiveness of grain size reduction during high-energy (HE) processing also varies depending on the material being processed. For pure metals, those with a higher melting temperature generally have smaller average grain sizes. On the other hand, alloys

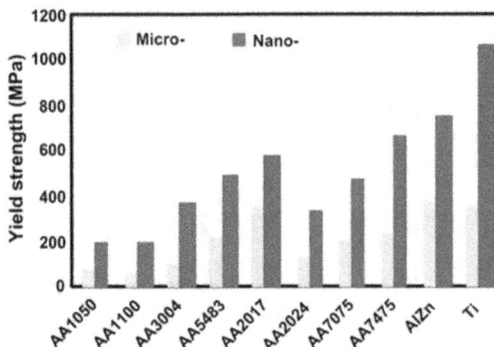

FIGURE 2.3   Yield strength of materials in micro- and nano-crystalline materials.

experience more significant grain size reduction during HE processing compared to pure metals. The homogeneity of grain size also plays a crucial role in determining the mechanical strength of nano-metals, as demonstrated by the variation coefficient (CV), which is defined as the ratio of standard deviation to the mean value of grain size. A decrease in the grain size homogeneity from CV = 0.07 to CV = 0.41 resulted in a decrease of about 100 MPa in yield strength for an average grain size of ~30 nm. However, nano-metals are more susceptible to non-homogeneity in grain size than conventional micro-grained metals, with a CV value often exceeding 0.45.

It has been found that the combined effect of grain refinement and precipitation hardening on the strength of a material is not a simple sum of their individual contributions [13]. This phenomenon is demonstrated in Figure 2.3, where it can be seen that for smaller grain sizes, the additional strengthening induced by precipitation is relatively small. This is due to the fact that the two mechanisms of strengthening are interdependent during processing. The presence of precipitates can make the process of grain refinement (e.g., through HE processing) more difficult, especially when the precipitates are small and coherent with the matrix. However, larger precipitates can actually facilitate grain size reduction [7]. In addition, the precipitation process in fine-grained materials is substantially different due to the high density of defects, such as dislocations and grain boundaries. As a result, the precipitates tend to form preferentially at these locations, leading to a different contribution to strengthening compared to coarse-grained structures.

Fatigue strength is a crucial mechanical property of engineering materials, and its enhancement in nano-metals has been a subject of significant interest. While nano-metals exhibit superior fatigue properties in the high cycle regime due to their high static strength, their low cycle fatigue properties are relatively weaker. This is mainly due to reduced ductility and a tendency to localize strain in the form of shear bands. As a result, the low cycle fatigue properties of nano-metals are accompanied by an increase in the rate of fatigue crack propagation. To explore this further, stress-controlled fatigue tests were conducted to obtain Wöhler curves for micro- and nanocrystalline samples (Figure 2.4). The results revealed a noticeable increase

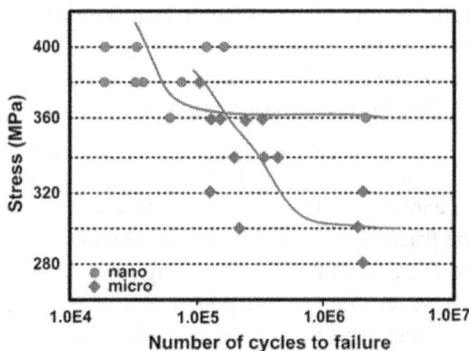

FIGURE 2.4 Wöhler curves for micro- and nano-crystalline alloy for different cycles of failure.

in the fatigue limit for nanostructured 2017 alloy, which was found to be 360 MPa compared to 295 MPa for the microcrystalline alloy. However, it should be noted that the improvement in tensile strength was much more significant, with the nanostructured samples exhibiting a tensile strength of 580 MPa compared to 310 MPa for the microcrystalline samples.

For microcrystalline materials, it is widely known that reducing the grain size not only increases their strength but also their ductility. This is achieved through the grain size strengthening mechanism, which enhances both the hardness and toughness of a material. By decreasing the grain size, crack deflection at grain junctions can occur, leading to increased constraint on crack propagation and thereby enhancing the fracture toughness and ductility of the material. However, when grain boundaries (GB)/triple junction migration becomes the dominant mechanism of plastic deformation, particularly as the grain size approaches the limit of instability of dislocations, early GB/triple junction crack interaction and crack propagation due to GB sliding can occur even at low temperatures, resulting in limited ductility. Although one would expect an increase in the ductility of nanocrystalline materials when the grain size is reduced to nanocrystalline dimensions, their ductility is often observed to be reduced compared to ultrafine-grained materials.

Koch identified three major sources of limited ductility in nanocrystalline materials, artifacts from processing (e.g., pores), tensile instability, and crack nucleation or shear instability. It is challenging to prepare bulk nanostructured materials free from such artifacts through consolidation, resulting in difficulty in synthesizing nanocrystalline bulk specimens suitable for tensile testing. Nanocrystalline and ultrafine-grained materials exhibit a different response to coarse-grained polycrystalline metal and cannot generally sustain uniform tensile elongation. After an initial stage of rapid strain hardening over a small plastic strain regime (1%–3%), the dislocation density saturates in nanocrystalline materials due to both dynamic recovery and annihilation at grain boundaries. Nevertheless, after large additional strains, work hardening is observed in these materials, while room-temperature dynamic recovery is also common in nanocrystalline samples. Finally, an increase and decrease in strain rate sensitivity with decreasing grain size in metals have been reported by different research groups. For instance, the strain rate exponent m of iron, which is normally strain rate sensitive and on the order of 0.04, decreases in value to 0.004 when the grain size is 80 nm.

Nanocrystalline materials exhibit different creep behavior than conventional grain-sized materials. The enhanced grain boundary area per unit volume in nanomaterials leads to higher reactivity and diffusivity, causing the faster formation of passive films. However, at lower stress and higher temperatures, the creep behavior of nanomaterials is generally faster due to the presence of high diffusivity paths. The increased volume fraction of grain boundaries in nanophase materials facilitates rapid diffusion along the grain interfaces, leading to enhanced creep rates even at room temperature. However, the level of porosity in nanophase samples can affect creep rates, as free surfaces tend to increase diffusion rates relative to grain boundary rates. Furthermore, doping nanocrystalline materials with certain elements, such as yttrium oxide, can prevent grain growth and lead to lower creep rates at temperatures lower than expected. The Ashby–Verral model based on grain boundary sliding with diffusional flow can be used to understand creep behavior in nanophase ceramics. When compressive stress is applied to a system of equiaxed grains, the grains slide over one another by diffusional accommodation at the interfaces, maintaining their size and orientation. Diffusion can also contribute to filling cracks opened at the interfaces. Overall, the creep behavior of nanomaterials is complex and cannot be solely attributed to enhanced grain boundary diffusion.

The reduction of grain size in nanometals has significant implications for their corrosion resistance. One consequence of this reduction is an increase in the surface area of grain boundaries per unit volume of the alloy. This contributes to higher reactivity and diffusivity, allowing passive films to form more quickly. Additionally, nano-refined metals tend to exhibit higher chemical homogeneity and a more uniform distribution of alloying elements, although second-phase inclusions may not be refined by severe plastic deformation (SPD) treatment. However, predicting the corrosion resistance of SPD-refined metals is not straightforward, as the results of corrosion resistance measurements can be ambiguous. For example, nano-titanium has been found to exhibit better corrosion resistance in various environments,

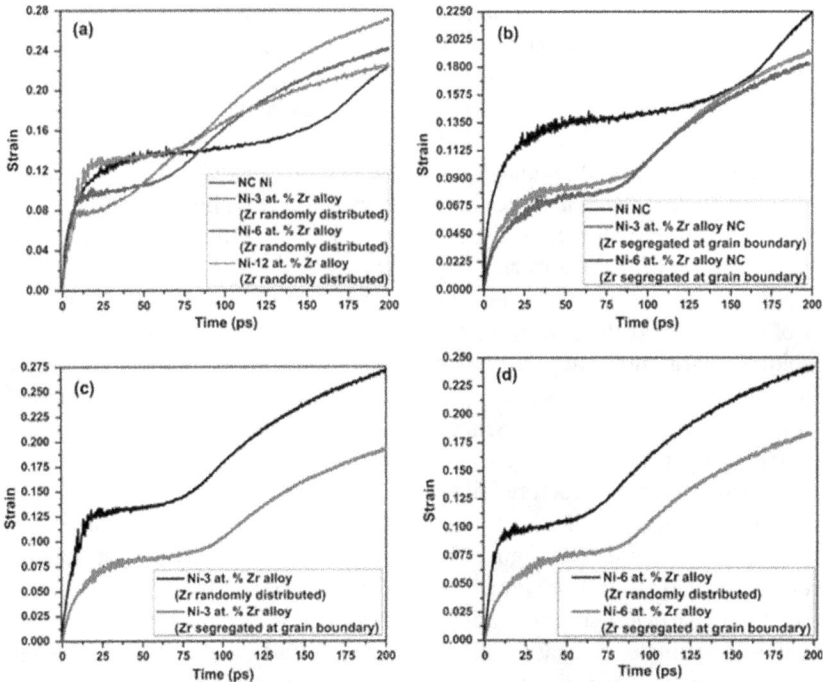

FIGURE 2.5 Simulated creep curves for nanocrystalline Ni specimen with various distributions of second element Zr at grain boundaries and in bulk: (a) Zr atoms randomly distributed; (b) Zr atoms segregated at GBs; (c) black line: Zr atoms randomly distributed, and red line: Zr atoms segregated at GBs; and (d) Ni-6 at. % Zr alloy (black line: Zr atoms randomly distributed, and red line: Zr atoms segregated at GBs) [Meraj and Pal, 2016].

including artificial saliva, simulated body fluid, and solutions containing F-ions. On the other hand, nano-AA7475 demonstrates less noble corrosion potential, higher passivation current, and a more negative breakdown potential, indicating a potential detrimental effect of grain size refinement on the corrosion resistance of this alloy. Further analysis of corroded surfaces and corrosion damage reveals that in the case of nano-AA7475 obtained via HE, corrosion attack concentrates only in the region around intermetallic particles. However, it is important to note that intergranular corrosion, which is known to reduce the lifetime of components made of this alloy in service conditions, does not occur in nano-AA7475. Overall, the relationship between grain size and corrosion resistance in nanometals is complex, and further research is needed to fully understand the effects of SPD refinement on their corrosion properties.

2.4   EVALUATION OF THERMAL PROPERTIES

The thermal conductivity of nanomaterials is subject to various factors, such as grain size, shape, and the presence of defects or dislocations. While increasing the number of grain boundaries tends to lower thermal conductivity by enhancing phonon scattering at disordered boundaries, nanocrystalline materials may exhibit distinct properties due to photon confinement and quantization effects. For instance, one-dimensional nanowires can have ultralow thermal conductivities resulting from the quantum confinement of phonons in 1D and strong phonon–phonon interactions. Silicon nanowires, for example, have thermal conductivity about two orders of magnitude smaller than that of bulk silicon. On the other hand, carbon nanotubes have a tubular structure that enables an extremely high axial thermal conductivity (~6600 W/mK). However, their heat transport property is highly anisotropic, making the thermal transport direction dependent. Additionally, in multilayered coatings, collective modes of phonon transport may appear, leading to significant changes in transport properties when the phonon coherence length is comparable to the thickness of each layer. If the multilayer has a superlattice structure with alternate films exhibiting a large mismatch in phonon dispersion relations, phonons in a certain frequency range may not propagate to neighboring layers unless there are mode conversions at the interface. The presence of interface dislocations and defects can also contribute to enhanced boundary scattering, further lowering the thermal conductivity of multilayered nanostructured films.

   Nanofluids, which are liquids containing a stable dispersion of nanoparticles uniformly distributed in the medium, offer a promising way to enhance thermal transport using the unique thermal properties of nanomaterials. The dispersion of a wide range of nanoparticles, such as oxides, nitrides, metals, metal carbides, and nanofibers, including single- and multi-walled carbon nanotubes, has been shown to significantly increase the thermal conductivity of fluids. In order to obtain stable colloidal suspensions, the particle size typically needs to be in the range of 1–100 nm, and an anti-coagulant may also be added to improve stability. The concept of enhancing the thermal conductivity of fluids by incorporating solid dispersions is not a new one. As early as 1873, Maxwell proposed the possibility of using particle dispersions to improve the thermal conductivity of fluids. However, in the past, microcrystalline dispersoid particles were commonly used, resulting in inferior suspension stability, which led to coagulation and precipitation. Additionally, the erosion of pipe walls by the particles was a major problem. With the development of techniques to synthesize nanoparticles with controlled grain size, nanofluids with improved stability have been developed, offering a promising avenue for enhancing thermal transport.

3 Nanoscale Simulation of Deformation Behavior

Nanoscale simulation of deformation behavior has become a vital tool for understanding the mechanical properties of materials at the atomic level. By employing computational methods, scientists and engineers can predict the deformation behavior of materials and design new materials with specific mechanical properties. The simulation of deformation behavior at the nanoscale involves nanoscale simulation by the use of molecular dynamics simulations. One of the key advantages of nanoscale simulation of deformation behavior is that it allows researchers to study the behavior of materials that are difficult or impossible to test experimentally. For example, it is challenging to measure the deformation behavior of materials at the nanoscale, where the effects of surface roughness and defects become increasingly significant. Another advantage of nanoscale simulation is that it can be used to study the effect of specific variables on the deformation behavior of materials. For example, researchers can use simulations to investigate the effect of temperature, strain rate, and defect density on the deformation behavior of materials. This information can be used to design materials with specific mechanical properties for specific applications. The use of nanoscale simulation of deformation behavior is becoming increasingly widespread in materials science and engineering. It has been used to investigate the deformation behavior of a wide range of materials, including metals, polymers, ceramics, and composites. As computational power continues to increase, and simulation methods become more sophisticated, the use of nanoscale simulation in materials science and engineering is expected to become even more prevalent. In conclusion, nanoscale simulation of deformation behavior is a powerful tool for understanding the mechanical properties of materials at the atomic level. By employing computational methods, researchers can predict the deformation behavior of materials under different loading conditions and investigate the effect of specific variables on the deformation behavior of materials. As computational methods continue to improve, nanoscale simulation is expected to become an even more essential tool for materials science and engineering research.

3.1 SCALE-DEPENDENT DEFORMATION BEHAVIOR

Nanostructured metals provide an excellent platform to investigate plasticity at different length scales. Multilayer systems consisting of nanoscale layers show remarkable strength and ductility, especially when the thickness of individual layers is just a few nanometers. Studies suggest that the strength of such systems can reach up to one-third of the theoretical strength when the thickness is around

 DOI: 10.1201/9781003323495-3

2–10 nm [Mara et al., 2010; Bhattacharyya et al., 2011; Han et al., 2009]. However, interpreting the plastic flow behavior of these systems using conventional nanoindentation hardness measurements is challenging due to the non-uniform stress state in the plastic zone.

On the other hand, researchers have extensively studied the mechanical fatigue behavior of metallic thin films and its dependence on length scales ranging from microns to nanometers. The results indicate that the fatigue behavior of thin films is considerably different from bulk materials due to the constraints imposed by their microstructural and geometric features [Judelewicz et al., 1994; Wang et al., 2008; Read, 1998]. At larger length scales, fatigue deformation occurs through the formation of extended dislocation structures accompanied by extrusions/intrusions, leading to weak length scale-dependent crack nucleation and propagation. However, at nanoscale lengths, the formation of extended dislocation structures is inhibited, and the accumulation of plastic strain within grains is restricted, resulting in the dominance of individual dislocation motion. As a result, interface-mediated damage and grain boundary processes become more prevalent in controlling the fatigue process.

The investigation of creep deformation in nanocrystalline metals is a challenging and costly task with limited research due to experimental constraints [Ashkenazy and Averback, 2012; Ghosh and Chokshi, 2014]. To address this issue, molecular dynamics simulations have emerged as a cost-effective and reliable approach to analyze creep deformation and elucidate underlying mechanisms in nanocrystalline materials [Desai et al., 2008; Berry et al., 2015]. For instance, Keblinski et al. [1998] employed MD simulations to demonstrate that Coble creep governs creep deformation in ultrafine-grained nanocrystalline materials ($d \leq 10$ nm). Additionally, researchers have conducted diffusional creep simulations to study the impact of stress-assisted grain boundary migration on diffusional creep [Trautt et al., 2012]. In nanocrystalline metals with ultrafine grain sizes, grain boundary diffusivity plays a vital role in creep behavior, with grain boundary diffusional creep being the dominant mechanism. MD simulations have been utilized to identify various factors influencing grain boundary diffusivity, including segregated solute at the grain boundary, grain boundary orientation, and surface atom diffusion. MD simulations provide a powerful tool to investigate deformation behavior at the atomic scale in nanocrystalline metals. The modeling of atomic interactions allows researchers to gain insights into the mechanisms governing the deformation behavior of these materials. Furthermore, MD simulations can validate experimental results and provide a deeper understanding of the underlying physics of deformation behavior. Comparing simulation results to experimental data can verify model accuracy and provide insights into factors contributing to differences between experimental and theoretical results. These simulations enable investigation of the behavior of nanocrystalline metals at the atomic scale under various conditions and can offer insights into the mechanisms responsible for their unique mechanical properties. This section presents an overview of scale-dependent deformation behavior, and subsequent sections will provide in-depth analysis through specific case studies.

Scale of Materials-Related Investigation

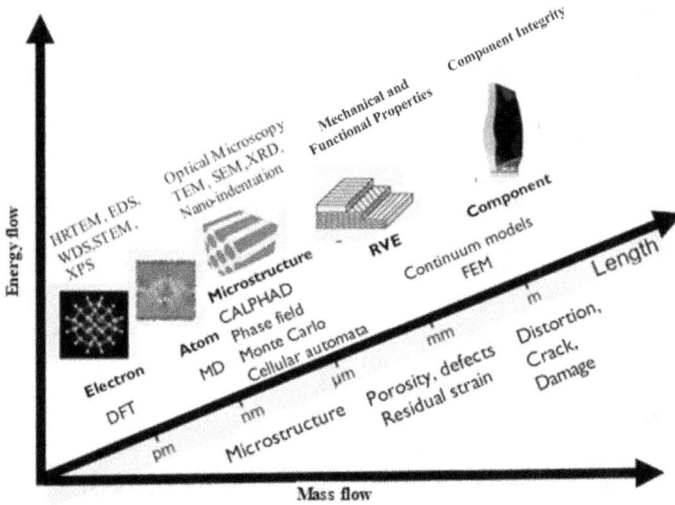

Capturing multi-scale structure-property correlation

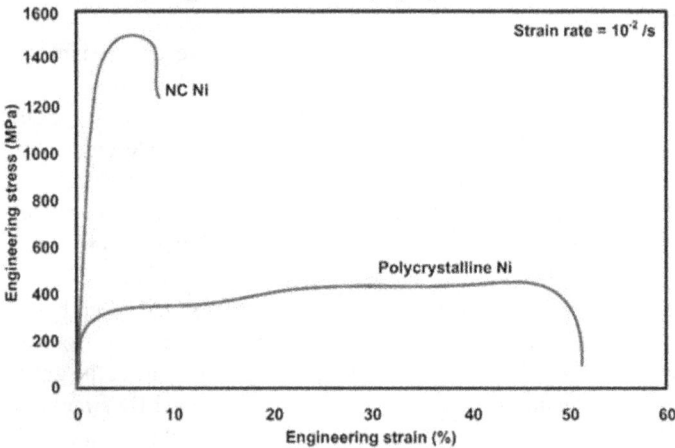

FIGURE 3.1 Spectrum of the length scale and its relation with the mass and energy flow, along with a detailed schematic of the computational models that can be implemented for each scale. The stress–strain curve shows the change in strength with the change in the scale of the specimen.

3.2 DEFORMATION SIMULATION OF DYNAMIC LOADING

3.2.1 NANOSCALE DEFORMATION MECHANISM: TENSILE AND COMPRESSION

The authors of this section conducted a study using molecular dynamics simulations to investigate the deformation behavior of Ni20W20Cu20Fe20Mo20 high-entropy alloy (20 at. % each element) single crystals under uniaxial tension-compression loading [Meraj and Pal, 2016]. The aim was to explore the impact of observed nano twins on deformation behavior for two different loading processes, tensile followed by compressive and compressive followed by tensile. To achieve this, classical MD simulations were performed using the Large-scale Atomic/Molecular Massively Parallel Simulator developed by the Sandia National Laboratory in the United States. The simulations were conducted at a temperature of $-10°C$ and a strain rate of 10^8 s^{-1}. To model the interactions between the Ni, W, Fe, Cu, and Mo atoms, the embedded atom method potential was utilized. The atomic positions were determined and visualized using the open visualization tool software OVITO, as shown in Figure 3.2. Overall, this study provides insights into the deformation behavior of high-entropy single-crystal alloys, with a focus on the influence of nano twins under different loading conditions.

Figure 3.3 shows the stress–strain curves obtained from the simulation of high-entropy alloy single crystals under uniaxial loading (both tension and compression)

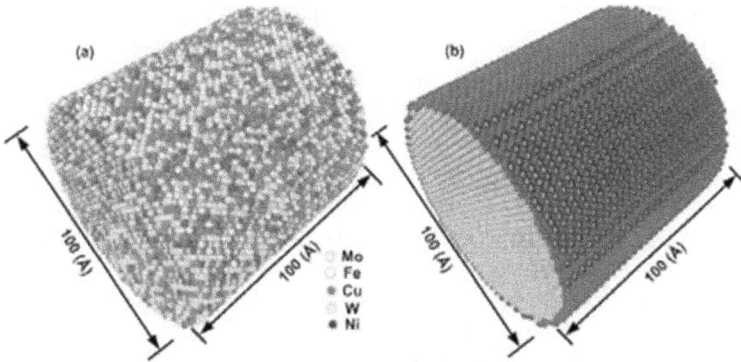

FIGURE 3.2 Atomic configuration of three-dimensional snapshots of the high-entropy alloy sample.

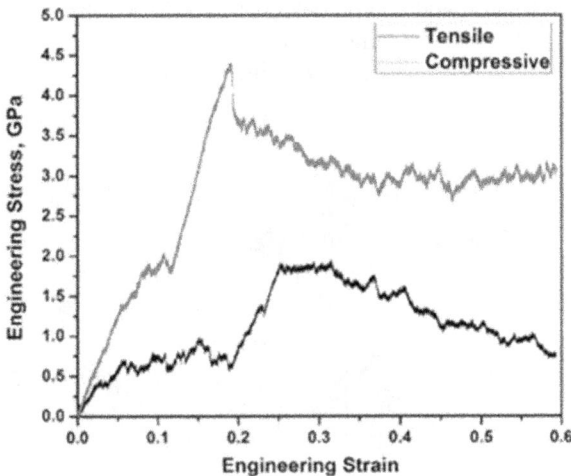

FIGURE 3.3 Engineering stress–strain curves for high-entropy alloy single crystals under uniaxial tensile and compressive loading separately.

at a temperature of $-10°C$ and a strain rate of 10^8 s^{-1}. The stress–strain curves exhibit four different stages, a linear elastic region, a small plastic region, another linear elastic region, and finally plastic deformation. The maximum tensile stress observed is around 1.9 GPa, corresponding to a strain of approximately 0.25. In contrast, the maximum compressive stress recorded is approximately 4.4 GPa, with a corresponding strain of approximately 0.19. The initial small plastic deformation observed in both tensile and compressive stress–strain curves is attributed to the formation of twins, whereas the subsequent linear–elastic portion of the curves corresponds to strain hardening.

On the other hand, Figure 3.4 presents the stress–strain curves of high-entropy alloy single crystals for tensile followed by compressive loading after a pre-strain of approximately 0.60, along with a pictorial representation of corresponding atomic snapshots colored according to CSP values at different positions of the deformation path. Figures 3.3 and 3.4 depict the stress–strain curves of high-entropy alloy single crystals under different loading conditions, along with snapshots of the atomic structure colored according to the centro-symmetry parameter values at various positions along the deformation path. For tensile loading followed by compression, plastic deformation is observed in the forward loading direction, with twinning observed at points a1 and a2 on the stress–strain curve. In contrast, a crystalline and amorphous structure is observed during the compressive loading phase after a strain of 0.6. Similarly, for compressive loading followed by tension, plastic deformation is observed in the forward loading direction, with twinning observed at point a1 on the stress–strain curve. Again, a crystalline and amorphous structure is observed during the compressive loading phase after a strain of 0.6. Linear elastic and serrated plastic regions are also observed. Notably, the yield stress is higher in compressive followed by tensile loading compared to tensile followed by compressive loading. Twins, which are inclined at approximately 45 degrees to the loading direction, are primarily generated during the forward loading phase (i.e., the tensile part of tensile followed by compressive and compressive part of compressive followed by tensile) and are observed to act as potential sources of amorphization. This observation is not reported in the available literature for single crystals subjected to these loading conditions.

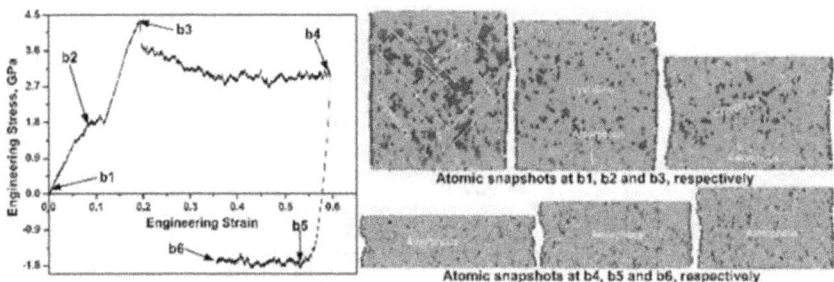

FIGURE 3.4 Stress–strain plots of high entropy alloy (HEA) single crystals for tensile followed by compressive loading after ~0.60 pre-strain, with the atomic snapshots marked on the corresponding point of the stress–strain curve.

3.2.2   UNIAXIAL LOADING USING LAMMPS

In this section, the focus will be on the investigation of the high-temperature deformation behavior of nanocrystalline Ni-Nb alloy with varying amounts of Nb (4, 8,
and 10 at. %) segregated at the grain boundary [Meraj and Pal, 2019]. Molecular
dynamics simulations were employed to perform tensile tests on the alloy at a temperature of 800 K and a strain rate of 10^8 s^{-1}. Structural changes under tensile loading
were analyzed along with dislocation density and vacancy formation. The results
demonstrate that the strength and flow stress of the nanocrystalline Ni-Nb alloy
specimens increases significantly with higher percentages of Nb atoms segregated
at the grain boundary. Here, we explore the high-temperature deformation behavior
of nanocrystalline (NC) Ni-Nb alloy with Nb solute segregated at grain boundaries.
The stress–strain behavior of NC Ni-Nb alloy specimens is represented in Figure 3.5.
It is observed that the stress–strain curves increase up to the ultimate tensile stress
and then gradually decrease. Furthermore, the stress–strain curves shift upwards
with an increasing percentage of Nb atoms at grain boundaries due to the effective
pinning of grain boundaries by the segregated solute at GBs, which suppresses the
grain boundary sliding motion. It is observed that the ultimate tensile stress (UTS)
gradually increases with increasing Nb solute at grain boundaries at 800 K temperature, as shown in Figure 3.5. This finding is similar to what was reported by Schäfer
and Albe [2012], who found that the strength of NC Cu is increased with increasing
Nb solutes at GBs, per MD simulation.

Figure 3.6 presents the atomic configuration snapshots for NC Ni-8 at. % Nb alloy
at 800 K temperature at four different stages of strain. Atoms are colored according
to the centro-symmetry parameter to show the structural changes that occur during
tensile deformation. The figure shows that various deformation mechanisms are activated during the tensile deformation of NC Ni-8 at. % Nb alloy, including intergranular crack nucleation, grain boundary sliding, and twin formation. It is noteworthy

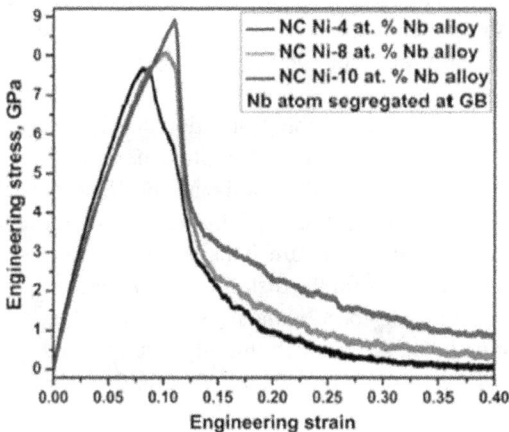

FIGURE 3.5 Engineering stress–strain plots of NC Ni-Nb alloy with different alloying
percentages (at. %) at 800 K temperature.

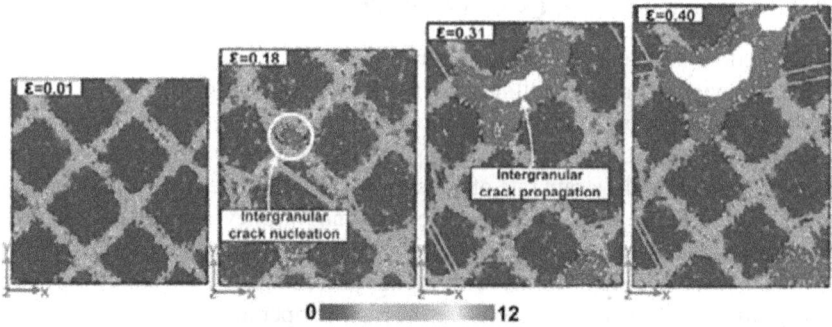

FIGURE 3.6 Atomic configuration snapshots of NC Ni-8 at. % Nb alloy specimen under high-temperature tensile loading at different strain values.

that intergranular cracks are found to nucleate at the intergranular region of NC Ni-8 at. % Nb alloy under tensile loading conditions, as observed in the figure. The intergranular crack nucleation strains vary for different NC Ni-Nb alloy specimens, with different at. % Nb segregated at grain boundaries at different strains. And it is observed that the intergranular crack nucleation strains increase with an increase in Nb atoms at grain boundaries. Furthermore, UTS and intergranular crack nucleation strain for NC Ni-Nb alloys are found to increase with increasing at. % Nb atoms at GB. These observations are consistent with previous studies that have reported an increase in strength with an increase in Nb solutes at GB per MD simulation, showing the reliability of MD simulation.

Another important dynamic analysis that can be performed with MD simulation is the time evolution of the dislocations and vacancies in the specimen during tensile deformation (in that case, any deformation). We present the results of an investigation into the behavior of dislocation density versus strain in nanocrystalline Ni-Nb alloys under high-temperature tensile loading conditions. The results are depicted in Figure 3.7, which shows that the dislocation density initially decreases with increasing strain up to a certain point for each alloy, after which it increases with further strain.

For the NC Ni-4 at. % Nb alloy, the dislocation density decreases up to approximately 0.10 strain. For the NC Ni-8 at. % Nb alloy, the dislocation density decreases up to approximately 0.16 strain. Finally, for the NC Ni-10 at. % Nb alloy, the dislocation density decreases up to approximately 0.19 strain. After these points, the dislocation density increases with further strain. This behavior is attributed to the presence of segregated solute atoms that pin the dislocations and restrict their movement from one grain to another through grain boundaries. The solute atoms effectively act as barriers to dislocation movement, which initially results in a decrease in dislocation density. However, as the strain continues to increase, the dislocations overcome the barriers, and the dislocation density increases. Another noteworthy observation from the results is that the dislocation density curves for each alloy are shifted downwards with increasing at. % Nb atoms at GBs under tensile loading conditions. This behavior is due to the increased superposition of dislocation movement with increasing

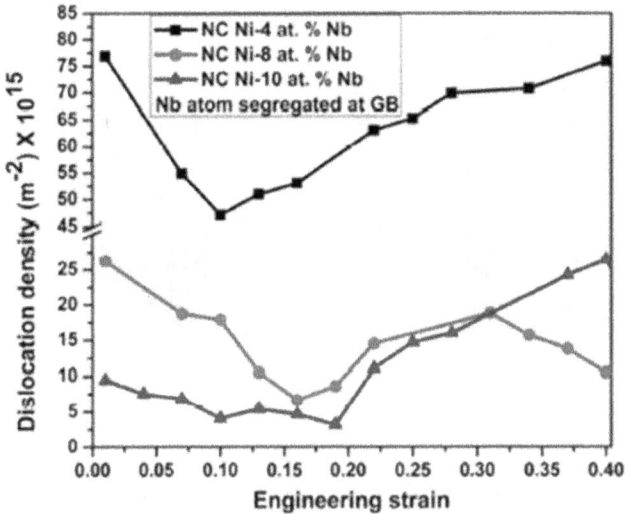

FIGURE 3.7 Dislocation density versus strain plots of NC Ni-Nb alloy at 800 K temperature showing the evolution and formation of dislocations during tensile loading condition.

Nb atoms at GBs. Overall, these results provide important insights into the behavior of dislocation density in NC Ni-Nb alloys under tensile loading conditions and may have implications for the development of high-strength materials.

3.2.3 BIAXIAL AND TRIAXIAL LOADING USING **LAMMPS**

In this section, we discuss the deformation behavior study of metallic systems under multiple loading conditions (from different directions), which is a more realistic deformation occurring to the materials around us. We present a molecular dynamics simulation study to investigate the mechanical behavior of nanocrystalline aluminum and carbon nanotube-reinforced NC Al composites under biaxial tensile loading (refer Figure 3.8 for structural configuration of the composite). The composites were subjected to tensile tests at a strain rate of 10^9 s^{-1} at three different temperatures (10 K, 300 K, and 681 K) using armchair-type (5,5) CNT, (15,15) CNT, and (30,30) CNT. The objective was to examine the effect of CNT on the mechanical properties of NC Al and to determine the influence of temperature and CNT diameter on the deformation behavior of the composites.

The simulation results showed that the incorporation of CNT in NC Al composites led to a reduction in ultimate tensile strengths and fracture strains compared to the NC Al without CNT specimen, corresponding to simulated temperatures. However, the (15,15) CNT specimen incorporating NC Al exhibited a higher fracture strain compared to the NC Al specimen at room temperature under biaxial loading, indicating an enhanced load-bearing capacity of the composite due to CNT reinforcement. Furthermore, the mechanical properties of the composites were found to decrease with increasing temperature due to the thermal fluctuations that occurred at elevated

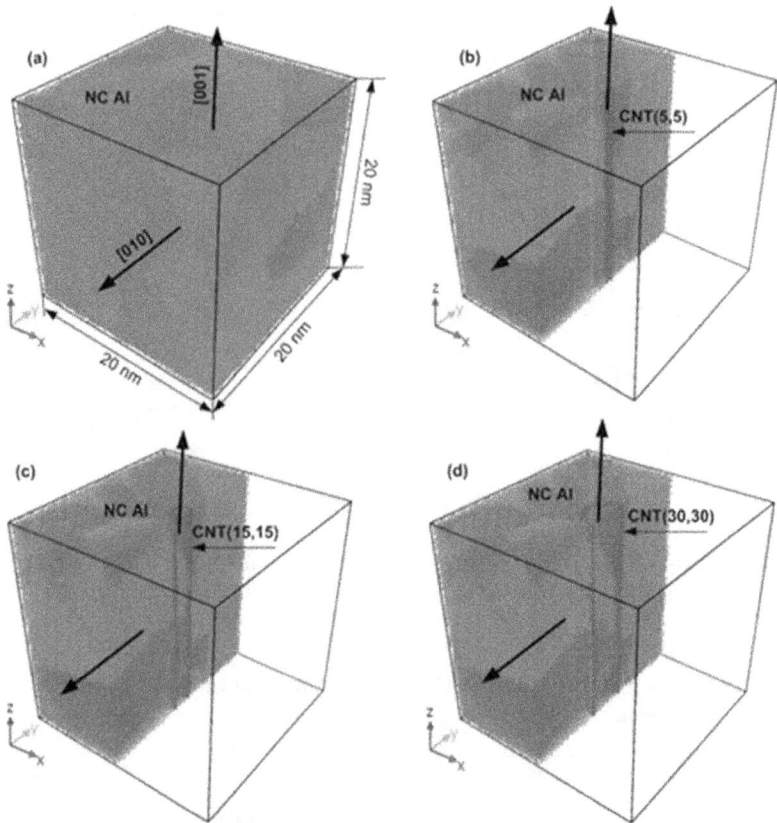

FIGURE 3.8 The initial configurations of the (a) NC Al without CNT and NC Al with different types of CNT, (b) (5,5), (c) (15,15), and (d) (30,30) during biaxial tensile loading condition along the y and z directions.

temperatures, which led to a reduction in the strength and ductility of the materials. Overall, our study provides insights into the mechanical behavior of CNT-reinforced NC Al composites under biaxial tensile loading, which can be useful in the design and development of high-performance materials for various applications.

Our results showed that the NC Al specimens exhibited higher ultimate tensile strengths compared to CNT embedded NC Al specimens at the same temperatures (Figure 3.9). The highest ultimate tensile strength was observed at 10 K, with a value of 7.24 GPa, while the lowest was observed at 681 K, with a value of 4.88 GPa. Except for the (15,15) CNT-reinforced NC Al specimen at room temperature, all other CNTs exhibited a reduction in fracture strains compared to NC Al specimens at the described temperatures. This reduction in fracture strain could be attributed to the load being transferred from the matrix to the reinforcement, leading to thermal fluctuations in CNT-reinforced NC Al specimens. Furthermore, our simulations showed that the mechanical properties of the NC Al and CNT-reinforced NC Al

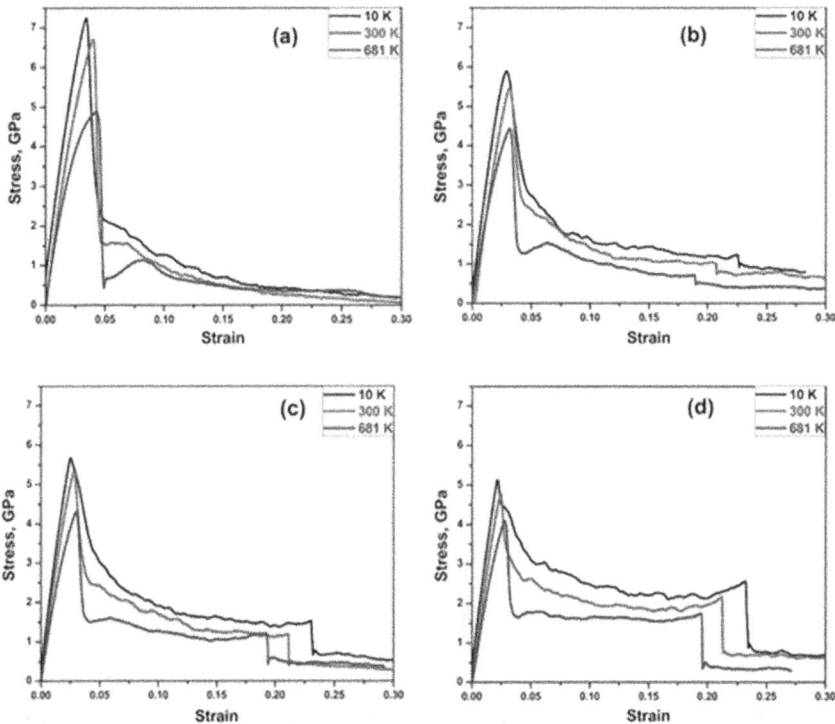

FIGURE 3.9 Stress–strain plots of (a) NC Al without CNT, (b) NC Al with CNT(5,5), (c) NC Al with CNT (15,15), and (d) NC Al with CNT (30,30) at different temperatures.

specimens decreased with increasing temperature. This decrease in mechanical properties was due to the thermal fluctuations that occurred at elevated temperatures, leading to a reduction in the strength and ductility of the materials. Under uniaxial tensile loading, our simulations revealed that the CNT-reinforced NC Al composites exhibited better enhancement in mechanical properties compared to NC Al specimens. However, under biaxial tensile loading, the CNTs did not influence the mechanical properties of the composites. This could be attributed to the directional dependence of CNTs, where they are stronger in the longitudinal direction compared to the radial direction. Overall, our study provides important insights into the mechanical behavior of CNT-reinforced NC Al composites under different loading conditions and temperatures, which can be useful in the design and development of high-performance materials for various applications.

3.3 DEFORMATION SIMULATION OF STATIC LOADING

3.3.1 Nanoscale Creep Deformation Process

Creep deformation is a time-dependent inelastic process that occurs under constant stress. It is a crucial factor that determines the durability and lifetime of many

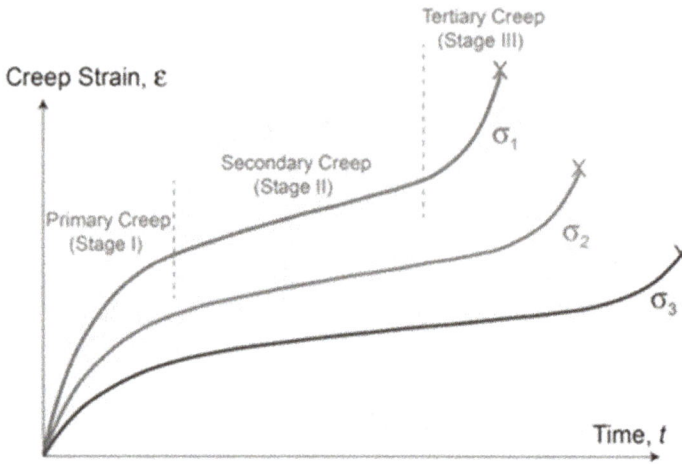

FIGURE 3.10   Creep deformation behavior at three different constant load values, showing the change in deformation behavior.

engineering materials. The creep process can be divided into three stages: primary, secondary, and tertiary creep, which eventually leads to fracture. Each stage of creep is associated with different mechanisms that drive the deformation process (Figure 3.10).

At the nanoscale, the behavior of materials under creep is influenced by their size and shape [Meyers et al., 2006]. The volume of the material decreases as the grain size decreases, which amplifies the relative impact of surface and grain boundary diffusion on the creep process, giving rise to the size effect. Furthermore, the surface area-to-volume ratio changes with the shape of the material, which can impact the diffusion process, leading to the shape effect. Therefore, understanding the influence of size and shape on diffusion is crucial to regulate creep at the nanoscale. Grain boundary diffusion, also known as Coble creep, governs the primary creep stage, involving the diffusion of atoms along grain boundaries resulting in creep deformation. The primary creep rate increases with decreasing grain size, proportional to d-3. In contrast, the secondary creep stage is determined by dislocation movements, causing the material to deform under constant stress, and remains constant irrespective of grain size, represented as d0. The tertiary creep stage is associated with lattice or bulk diffusion, known as Nabarro-Herring creep, wherein atoms diffuse through the bulk of the material. The tertiary creep rate decreases with decreasing grain size and is proportional to d-2. In essence, the creep behavior of nanoscale materials is primarily influenced by their size and shape, which directly affect the diffusion mechanisms underlying deformation. To regulate creep at the nanoscale, it is imperative to explore the impact of size and shape on diffusion and devise strategies to manipulate these factors in engineering materials.

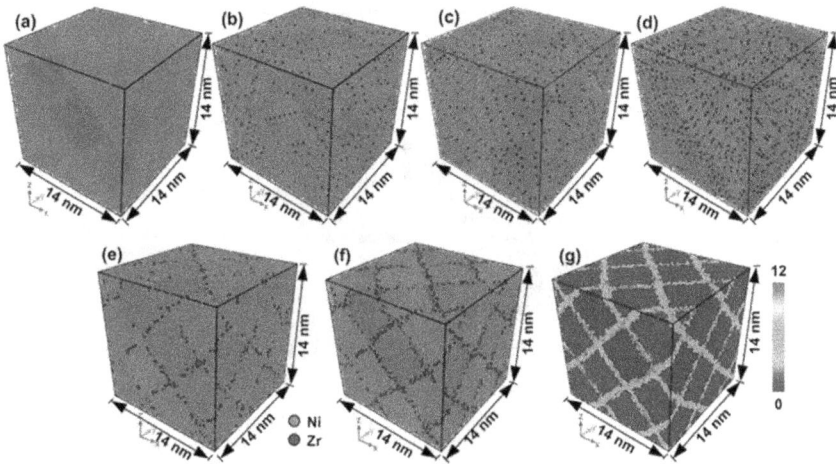

FIGURE 3.11 Three-dimensional NC Ni specimens highlighting randomly segregated Zr and Zr segregation at GBs. Also, the percentage of the secondary particle is varied to 3 at. %, 6 at. %, and 12 at. %.

3.3.2 Molecular Dynamics of Normal Creep Deformation

In this section, the authors present a case study that utilizes molecular dynamics simulations to investigate the creep behavior of nanocrystalline Ni and Ni-Zr alloys with a grain size of approximately 6 nm (Figure 3.11). These materials are of interest for high-strength, high-temperature applications, but their creep behavior is not well understood. The simulations were performed at a constant high temperature of 1209 K and a constant applied load of 1.0 GPa, and the effects of Zr additions (3 at.%, 6 at.%, and 12 at.%) were considered. The Zr was both randomly distributed in the NC Ni sample and segregated at the grain boundaries to evaluate its effect on the creep behavior. To model the atomic interactions, the embedded atom method potential was used. Overall, this case study demonstrates the utility of MD simulations for investigating the creep behavior of nanocrystalline materials, and the findings can inform the development of strategies for controlling creep at the nanoscale in engineering applications [Pal et al., 2017].

Figure 3.12 presents the creep curves obtained from molecular dynamics simulations of NC Ni alloys with varying atomic percentages of Zr addition. The results indicate that as the atomic percentage of Zr increases, the creep curves shift towards lower strain, with the exception of the NC Ni-3 at.% Zr alloy, where Zr is randomly distributed. This deviation is attributed to the low probability of Zr atoms being present at the grain boundary and causing disturbances in the atomic arrangement within the grains. The primary and secondary creep regimes of NC Ni and Ni-Zr alloys with randomly distributed Zr atoms are similar, but a significant difference is observed in the secondary creep regime of NC Ni and Ni-Zr alloy specimens. NC Ni exhibits a longer secondary creep regime than Ni-Zr alloys with randomly distributed Zr atoms. This is due to the earlier initiation of transformation to the amorphous structure for

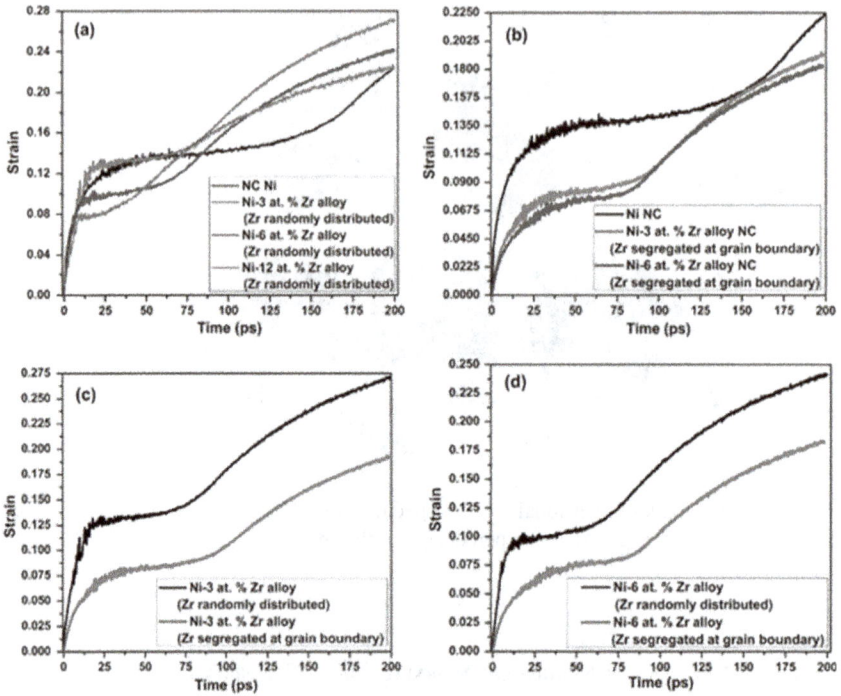

FIGURE 3.12 Simulated creep curves for different systems of Ni-Zr alloy. The black line represents randomly distributed Zr atoms, and the red line represents Zr atoms segregated at GBs.

NC Ni-Zr alloys with randomly distributed Zr atoms. The influence of segregated Zr atoms at GBs on creep behavior is found to be more significant compared to randomly distributed Zr atoms in the specimen. The study also analyzed the creep strain rate vs. time curves for NC Ni and Ni-Zr alloys with different atomic percentages of Zr addition. Initially, the creep rate decreased up to a certain time, followed by a constant creep rate region. After that, the creep rate increased and then decreased again after a certain time. The increase in creep rate is attributed to the disturbance of the regular atomic arrangement in the specimen, while the decrease in creep rate in the later stage is due to the presence of a formed amorphous structure. At this stage, negative creep becomes operative, which usually occurs when the internal stress of materials becomes higher than the applied external stress. Overall, this case study demonstrates the potential of MD simulations to investigate the deformation behavior of nanocrystalline materials and the influence of solutes on creep behavior.

Elevated temperatures can significantly impact creep deformation by causing the formation of point and line defects during the deformation process. In this study, we aim to analyze the behavior of point defects, specifically vacancies, during the creep deformation process in metallic systems. The Wigner–Seitz cell method was used to identify vacancies by comparing the defective crystal with a perfect crystal lattice. Interestingly, the number of generated vacancies initially increases, decreases, and

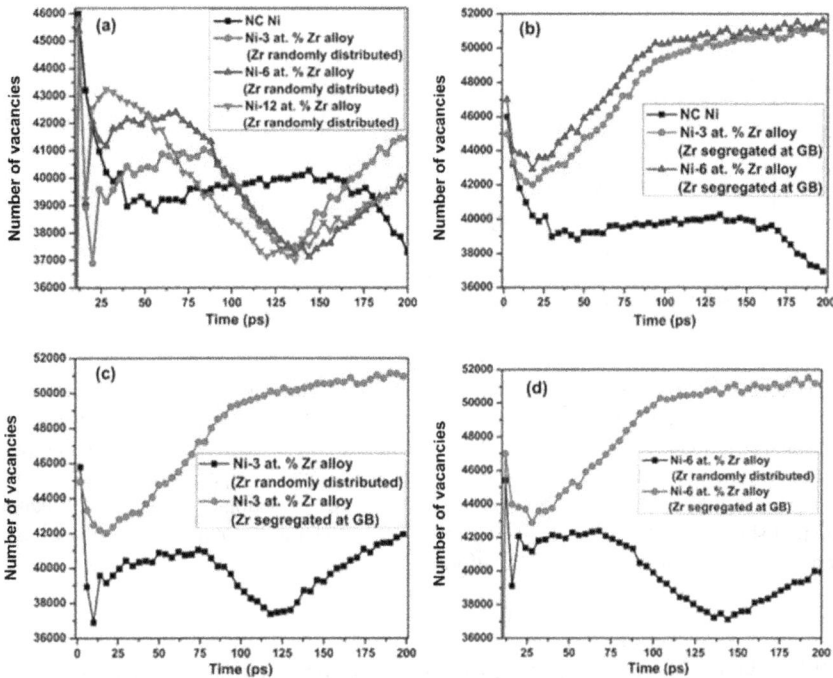

FIGURE 3.13 Plots of the number of formed vacancies as a function of time for different Ni-Zr systems. The red line corresponds to the Ni matrix with randomly distributed Zr, while the black line represents the Ni-Zr system with Zr segregated at grain boundaries.

almost becomes zero at around 135 picoseconds (ps) for NC Ni. However, for an NC Ni-Zr alloy with randomly distributed Zr atoms, the number of vacancies generated increased again with the progress of creep deformation. Additionally, the effect of Zr atoms segregated at grain boundaries was investigated, and the results showed that the number of vacancies generated decreased initially and then increased monotonically with the progress of the creep process for Ni-Zr alloy with 3 at. % and 6 at. % Zr atoms segregated at GBs (Figure 3.13).

It is worth mentioning that the number of vacancies generated for NC Ni decreased initially, went into a steady state up to around 160 ps, and then decreased during the creep process. On the other hand, the number of vacancies generated for Ni-Zr alloy with Zr atoms segregated at GBs was much more than for the other specimens, which was attributed to the consumption of the total excess volume of GBs. This led to an increase in grain boundary disorder, which enhanced hindrance to atomic movement across the grain boundary, contributing to the better creep properties exhibited by the Ni-Zr alloy with Zr atoms segregated at GBs. These findings highlight the significance of understanding the role of point defects, particularly vacancies, in determining the properties and behavior of materials. Moreover, they underscore the potential benefits of introducing specific point defects, such as Zr atoms segregated at GBs, to enhance the performance of materials in specific applications.

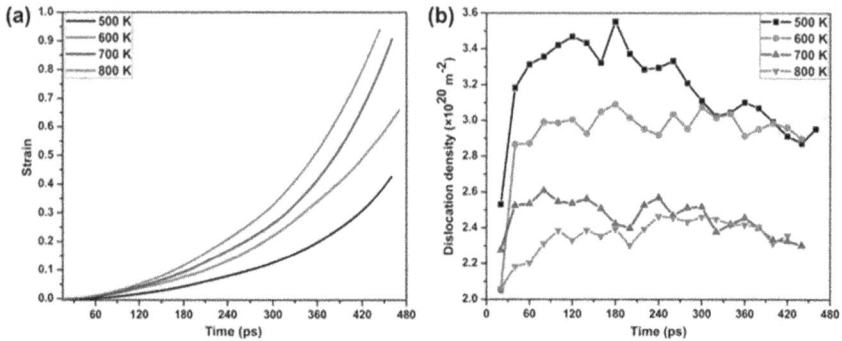

FIGURE 3.14 Deformation behavior of a nanocrystalline Ni nanowire specimen under constant load bending tests at different temperatures. (a) Plot of strain vs. time for nanowire specimen deformed at 500 K, 600 K, 700 K, and 800 K. The strain rate increases with time for all temperatures, and the specimen fractures rapidly at higher temperatures, except at 500 K. (b) Variation in dislocation density with time during the deformation of the nanowire specimen. The plot shows that dislocation density initially increases rapidly, reaches a steady state, and then decreases towards the end of the deformation.

3.3.3 MOLECULAR DYNAMICS SIMULATION OF BENDING CREEP PROCESS

Another important creep deformation process is bending creep, which exerts a multi-axial constant loading condition similar to the biaxial tensile loading condition discussed previously. However, the deformation behavior under these static conditions can be entirely different from the dynamic conditions. In this section, MD simulations are implemented to understand various deformation processes during the bending creep process of nanocrystalline material. This section presents a case study of molecular dynamics simulation study investigating the deformation and fracture behavior of a nanocrystalline nickel (Ni) nanowire specimen subjected to constant load bending tests at various deformation temperatures [Reddy and Pal, 2018a].

The study aimed to investigate the deformation behavior of a nanocrystalline Ni nanowire specimen through constant load bending tests performed at different temperatures (Figure 3.14). The results showed that the strain rate increased with an increase in the time period for all temperatures, which is consistent with the previous literature on nanocrystalline materials. During the bending deformation, the nanowire specimen was found to fracture rapidly at higher temperatures, except for 500 K, where an early fracture occurred. This is due to a higher plastic deformation caused by grain movement and lattice diffusion process. The increase in temperature causes a higher plastic deformation due to grain movement and lattice diffusion process, which is consistent with findings in previous literature. Furthermore, the corresponding dislocation density increased rapidly and attained a steady state, followed by a decrease at the end of the deformation. This suggests that dislocation-mediated deformation is the prevailing mechanism at lower temperatures, while the diffusion-mediated deformation process becomes more active at higher temperatures. Figure 3.15 shows that the maximum strain attained before failure increased with increasing temperature,

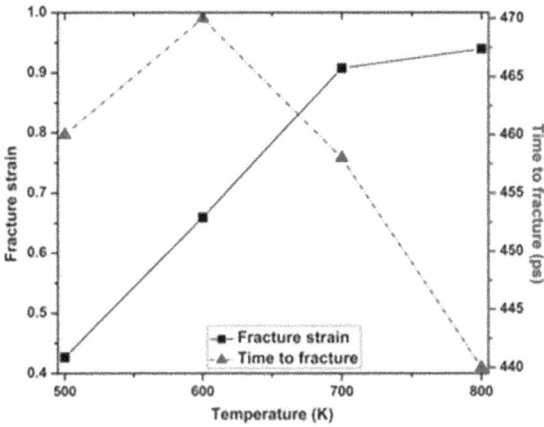

FIGURE 3.15   Plots present the variation in fracture strain and time of fracture with respect to deformation temperatures during the constant load bending deformation of a nanocrystalline Ni nanowire specimen.

with the highest fracture strain of 93.9% observed at 800 K. The fracture time was also observed to decrease with increasing temperature, with the shortest time recorded at 440 ps for 800 K. The failure of the specimen deformed at 500 K occurred due to brittle fracture, while the nanowire at 800 K underwent plastic deformation before fracturing due to the highest fracture strain observed. In summary, the study revealed that the deformation behavior of nanocrystalline Ni nanowire specimens is temperature dependent, with higher plastic deformation occurring at higher temperatures. The prevailing deformation mechanism was found to be dislocation-mediated deformation at lower temperatures and diffusion-mediated deformation process at higher temperatures. The results of the study could have significant implications in the design and development of nanocrystalline materials for various applications.

Additionally, we also present the atomic strain snapshots that were obtained to analyze the deformation mechanisms involved in the bending process. It was observed that the bending deformation occurred through slip and twinning mechanisms, and the failure of the specimen occurred by means of cleavage fracture. At an initial time period of 60 ps, a small magnitude of shear strain was generated along the grain boundaries at the lower portion of the specimen. As the time period increased, the shear strain in the specimen increased along the grain boundary, with the upper portion experiencing higher shear strain compared to the lower portion. Slip bands were found to be generated in the nanocrystalline specimen, either from the surface of the specimen or from one end of the grain boundary to the other end in the grain interior. The slip bands present in different grains had different orientations, which is reported in the literature. Shear bands were also formed in non-crystallographic orientations and could traverse through interfaces from one grain to another. The phenomenon of formation and propagation of the shear band during the constant load bending deformation process was studied, and it was observed that shear band formation occurred during the slip band and twin interaction (Figure 3.16). The presence of twin boundaries near the bent portion of the specimen contributed to the formation of shear bands due to slip-twin interaction. On further increase in time, the shear band propagated along the specimen, and finally, the failure of the specimen occurred through cleavage fracture. This analysis provides insights into the deformation and failure behavior of NC Ni nanowires under bending deformation and contributes to the development of better nanomaterials for various applications [Reddy and Pal, 2018b].

We further investigate the fracture mechanism of nanocrystalline Ni nanowires during the constant load bending deformation at 500 K. Figure 3.17 shows the CNA snapshot of the cross-sectioned specimen at a time period of 400 ps and the

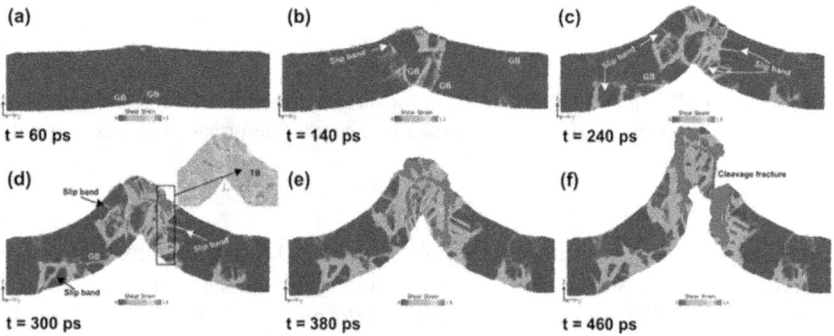

FIGURE 3.16   Atomic strain snapshots of the NC Ni nanowire specimen at 500 K. (inset) CNA snapshot of the bent section at a time period of 300 ps showing the presence of twin boundaries.

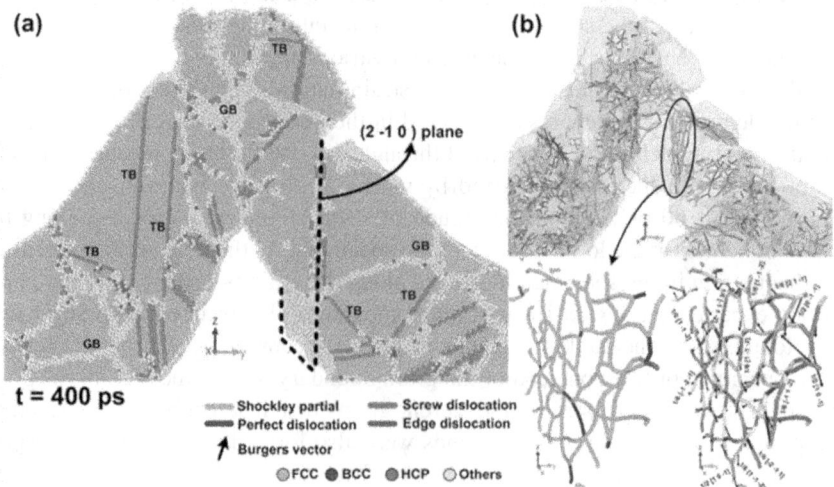

FIGURE 3.17   (a) Atomic snapshot of the bent section of the specimen deformed at 500 K. (b) Corresponding snapshot of the distribution of perfect and partial dislocations.

corresponding snapshot of the various partial and perfect dislocations generated in the specimen. It was previously observed that the cleavage fracture was initiated due to the shear band formation along the (2–1 0) plane. The dislocation analysis shows that a network of 1/6<1 1 2>Shockley partial screw dislocations is present at the right portion ((2–1 0) plane) of the bent specimen, and twin boundaries are also observed present at the exact location of the specimen. This indicates that the presence of twin boundaries is responsible for the stacking of the partial dislocations, and similar findings are also reported by other researchers. Twin boundary-slip band interactions experience significant shear strain and are known to promote crack initiation in nanocrystalline specimens due to the stacking of partial dislocations on the twin boundary plane. Hence, it is clear from this study that the specimen failed due to cleavage fracture, which was initiated by the crack formation along the (2–1 0) plane during the bending deformation occurring at 500 K. These findings provide valuable insights into the fracture mechanism of NC Ni nanowires and can aid in the development of better nanomaterials with improved fracture resistance for various applications.

The MD simulation can also be extended to very high temperatures, where creep deformation occurs extremely rapidly in nanoscale materials. In the case of bending creep deformation at 1300 K in bi-crystal Ni, the creep curve shows a steep increase in strain without a distinct steady-state creep region before fracture [Reddy et al., 2017]. Figure 3.18 provides a snapshot of the fracture mechanism during the bending deformation of a specimen at 1300 K. Unlike in the case of bending deformation at lower temperatures, the dislocation generation is not very prominent in the specimen,

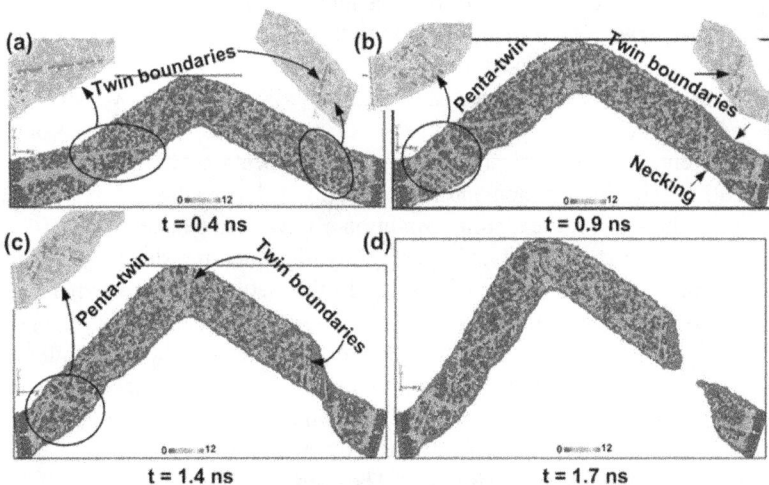

FIGURE 3.18 Fracture mechanism of bi-crystal Ni during bending creep at 1300 K. (a) Creation of twin boundaries at the ends of the specimen. (b) Formation of penta-twin occurs at the left end of the specimen, and necking is initiated at the right end of the specimen. (c) Cross-section at the right end region of the specimen decreases due to necking. (d) The penta-twin has completely disintegrated.

as the bending creep deformation at higher temperatures is a diffusion-mediated creep process. Figure 3.18(a) shows that grain-boundary migration has already occurred at 0.4 ns, and twin boundaries are generated at both ends of the specimen. These twin boundaries help in the high plastic deformation of the specimen. After a time period of 0.9 ns, the occurrence of a penta-twin is observed at the left end of the specimen, which is also illustrated by means of a CNA snapshot (inset) in Figure 3.18(b). The presence of the penta-twin on the left side of the specimen enhances the plasticity, ductility, and strength in that region. Hence, it can be seen that necking is initiated at the right end of the specimen. After a time period of 1.4 ns, it is observed in Figure 3.18(c) that the penta-twin starts to disintegrate due to lattice diffusion. The cross-sectional area of the right side of the specimen is substantially decreased, as seen in Figure 3.18(c), and subsequently, the fracture occurs at 1.7 ns, as shown in Figure 3.18(d). This study provides valuable insights into the fracture mechanism during high-temperature bending creep deformation, highlighting the role of diffusion-mediated creep and twin boundaries in enhancing the plasticity and ductility of the material.

3.4 DEFORMATION SIMULATION OF IMPACT AND CYCLIC LOADING

3.4.1 SHOCK DEFORMATION IN NANOSCALE METALLIC SYSTEMS

Shock deformation in nanoscale metallic systems is a phenomenon that has received increasing attention in recent years due to its potential applications in various fields such as material science, engineering, and manufacturing. This process involves subjecting metallic systems to high strain rates and pressures, resulting in rapid deformation and a change in their mechanical properties. Shock deformation in nanoscale metallic systems can be achieved through different methods, including high-pressure shock wave loading, laser-induced shock compression, and high-velocity impact. These methods generate high-pressure shock waves that propagate through the metallic system, leading to plastic deformation, dislocation motion, and eventually fragmentation. The deformation and fragmentation behavior depend on the material properties of the system, the loading conditions, and the size of the system.

A common experimental setup for generating high-pressure shock waves involves the use of diamond anvils and a drive laser. In this setup, a small sample is placed between two opposing diamond anvils, which serve as the pressure-generating medium (Figure 3.19). The drive laser is then used to create a shock wave that propagates through the diamond anvils and into the sample. To visualize this setup, one can imagine a schematic cross-section of the diamond anvils and sample. The diamond anvils are typically shaped like truncated octahedrons, with flat, polished faces that come into contact with the sample. The sample is placed between these two anvils, typically in the form of a thin disc or plate, and is held in place by a gasket made of a high-strength material such as steel. The drive laser is typically positioned to the left of the diamond anvils, with its beam directed toward the sample. When the laser is fired, it generates a high-intensity pulse of light that rapidly heats and vaporizes a small region of the sample. This sudden expansion creates a shock wave

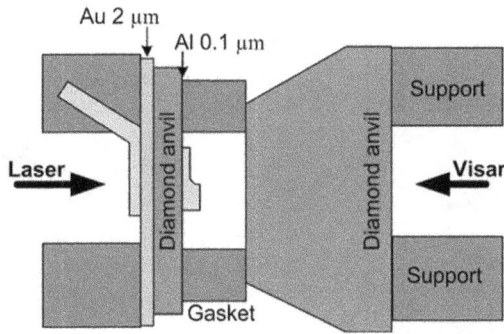

FIGURE 3.19 A experimental setup for generating high-pressure shock waves showing diamond anvils and a drive laser. The sample is placed between two opposing diamond anvils, and the drive laser is used to generate a shock wave that propagates through the diamond anvils and into the sample.

that propagates through the sample and into the diamond anvils, generating high pressures and temperatures. The shock wave can be detected and measured using a variety of experimental techniques, including optical microscopy, X-ray diffraction, and Raman spectroscopy. These techniques allow researchers to study the deformation behavior and phase transformations that occur in the sample under extreme pressure and temperature conditions.

The shock deformation process in nanoscale metallic systems can also result in the generation of high-density defects, such as dislocations, vacancies, and stacking faults, which can significantly affect the material's properties. High-density defects can lead to increased hardness, improved wear resistance, and enhanced mechanical strength. Moreover, the shock deformation process in nanoscale metallic systems can also be used to produce nanocrystalline materials, which exhibit unique mechanical and physical properties compared to their bulk counterparts. Nanocrystalline materials can have increased strength, improved ductility, and enhanced fatigue resistance, making them suitable for various applications.

Another important aspect of shock deformation in nanoscale metallic systems is the effect of strain rate on mechanical properties. The high strain rates associated with shock deformation can cause significant changes in the deformation behavior and the material's response to external loading. This effect is particularly pronounced in nanoscale metallic systems due to their small size and high surface area-to-volume ratio. Recent studies have shown that the strain rate sensitivity of nanoscale metallic systems can be significantly enhanced under shock deformation conditions. This enhanced sensitivity can lead to increased strain hardening and improved ductility, making nanoscale metallic systems ideal for high-performance applications that require high strength and toughness. Moreover, the shock deformation process can also induce phase transformations in nanoscale metallic systems [Reddy et al., 2019]. The high-pressure shock waves generated during the deformation process can cause the material to undergo a phase change, resulting in the formation of new crystal structures with unique mechanical properties (Figure 3.20). This phase transformation behavior

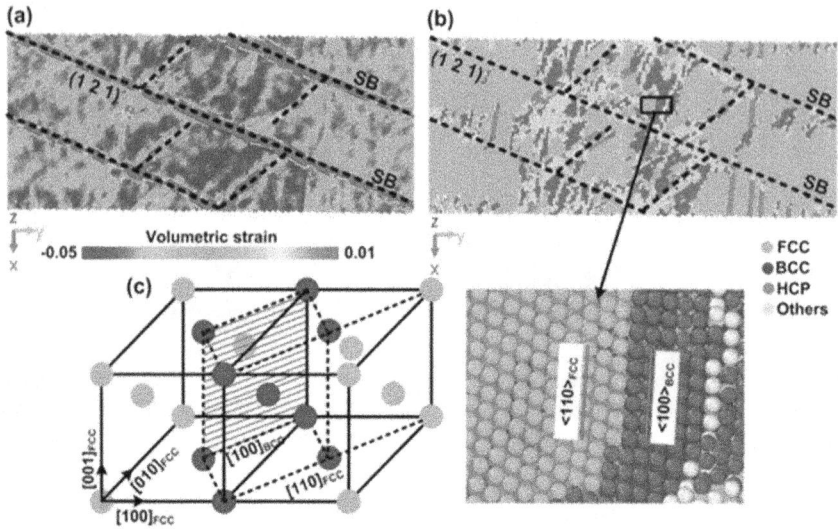

FIGURE 3.20 Simulated snapshot shows strain-induced phase transformation from FCC crystal structure to BCC structure during the shock propagation

is particularly important in the development of advanced materials with tailored properties for specific applications.

One of the challenges associated with shock deformation in nanoscale metallic systems is the difficulty in characterizing the deformation behavior and the resulting changes in the material's properties. The rapid deformation and fragmentation behavior can make it challenging to obtain accurate measurements of the material's mechanical properties, such as hardness, strength, and ductility. However, recent advances in experimental and simulation techniques, such as in-situ microscopy, high-speed imaging, and molecular dynamics simulations, have enabled more precise characterization of the deformation behavior in nanoscale metallic systems. In conclusion, shock deformation in nanoscale metallic systems is a fascinating area of research that has the potential to revolutionize the development of advanced materials with tailored properties for various applications. The ability to control and manipulate the deformation behavior and the resulting changes in the material's properties can lead to significant improvements in performance and reliability in high-performance applications.

3.4.2   Cyclic Loading Condition in Metals

Cyclic loading is a type of mechanical loading that involves the repeated application of stress or strain to a material. This loading condition is common in many engineering applications, such as in the design of aircraft, bridges, and automotive components, where materials are subjected to repeated loading and unloading cycles during operation. In metals, cyclic loading can lead to fatigue failure, which occurs when a material fails due to repeated stress cycles that cause small cracks to propagate and

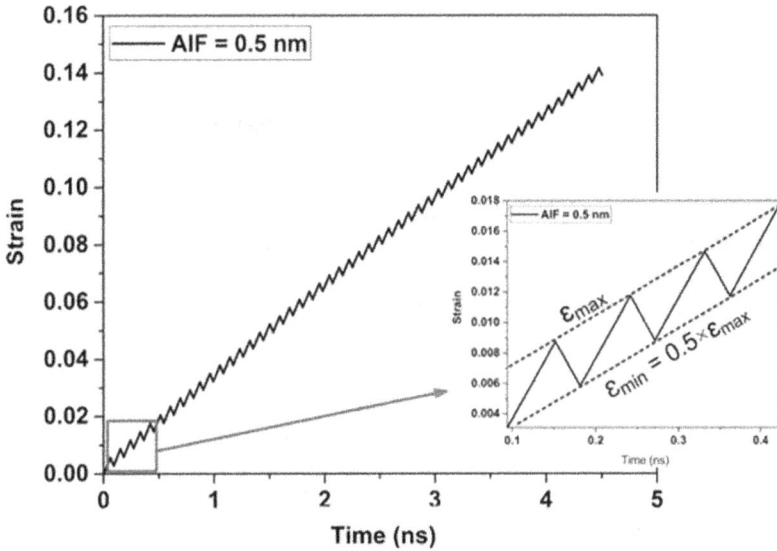

FIGURE 3.21 Strain vs. time plot during the cyclic loading of the specimen having different structural configurations for a total of 50 cycles with a load ratio, R of 0.5.

eventually coalesce, resulting in catastrophic failure (Figure 3.21) [Pal et al., 2019]. The fatigue life of a material is determined by the number of cycles it can withstand before failure occurs, and this life can be affected by a range of factors, including stress amplitude, mean stress, frequency, and environmental conditions.

One of the key challenges in understanding cyclic loading behavior in metals is the complex interaction between microstructure, deformation mechanisms, and crack propagation. Under cyclic loading conditions, metals undergo a range of deformation mechanisms, including slip, twinning, and dislocation movement, which can lead to changes in the microstructure and the development of localized plastic deformation. To study cyclic loading behavior in metals, researchers use a range of experimental techniques, including fatigue testing machines and microscopy techniques such as scanning electron microscopy (SEM) and transmission electron microscopy (TEM). These techniques allow researchers to observe the deformation behavior and crack propagation mechanisms in real time and to develop a better understanding of the underlying mechanisms that govern fatigue failure. However, to understand the cyclic load behavior at dynamic conditions and for nanocrystalline specimens, atomistic simulations are more suited. On the other hand, to improve the fatigue life of metals under cyclic loading conditions, researchers and engineers can use a range of techniques, including improving parameters such as the material's microstructure, reducing the stress amplitude, and controlling the loading frequency. By understanding the underlying mechanisms that govern cyclic loading behavior in metals and developing effective strategies to mitigate fatigue failure, researchers and engineers can design more reliable and durable materials for a range of applications. The next section will demonstrate the use of MD simulation in designing and analyzing the shock and cyclic loading conditions on NC specimens.

3.4.3   MODELING IMPACT AND CYCLIC LOADING CONDITION USING LAMMPS

In this section, we investigate the dynamic response of Cu-amorphous $Cu_{63}Zr_{37}$ nanolaminates under shock loading (taken as a case study), with a focus on understanding the overall deformation behavior of the nanolaminates in relation to different grain structures in the crystalline region (Figure 3.22). Pressure profiles of the single crystalline Cu-$Cu_{63}Zr_{37}$ metallic glass (SC/MG) nanolaminate at relatively low shock velocity indicate the presence of an elastic precursor in the crystalline region due to the plane–plane collision phenomenon. As shock velocity increases, the SC/MG specimen undergoes an FCC to BCC phase transition in the crystalline region, with the crystalline/amorphous interface generating a reflected rarefaction wave back into the crystalline region that aids in the evolution and stabilization of the BCC phase.

The present study aimed to investigate the pressure distribution dependency on shock velocities at different time steps in order to gain insights into the dynamic behavior of crystalline Cu-amorphous $Cu_{63}Zr_{37}$ nanolaminates. To achieve this, pressure profiles of the specimens with respect to distance were plotted during the shock propagation for different piston velocities. Figures 3.23(a–d) illustrate the variation in the compressive pressure in the SC/MG specimen for piston velocities of 0.5, 0.8, 1.1, and 1.4 km/s, respectively, during the shock loading from the crystalline region. The results showed that at slower piston velocity (i.e., 0.5 km/s), elastic precursors were formed in the crystalline region at an initial time period (i.e., 2.5 ps), as evident from the trend of pressure change in Figure 3.23(a). The oscillating pressure profile indicated the compression and release process similar to a harmonic oscillator. This oscillating behavior was due to the plane–plane collision that occurred during the shock propagation. The distance between the crest and trough of the oscillation was 6 Å, which is also the approximate thickness of the piston. However, as the piston velocity increased, the alternate increase and decrease nature of the pressure profile vanished, indicating that plastic deformation had dominated elastic deformation. This resulted in amorphization and phase transition, which will be discussed in more detail in a later section. With the propagation of shock towards the crystalline-amorphous interface (i.e., at 5 ps and a distance of ~25 nm in Figure 3.23(a)), it was found that the pressure profile showed a dip near the crystalline region and a peak

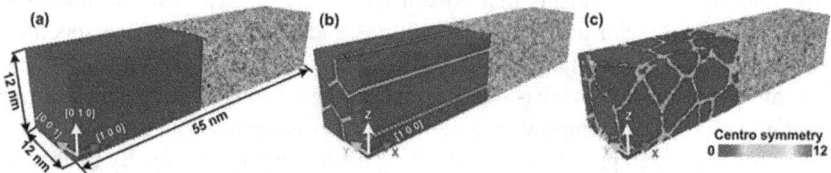

FIGURE 3.22 Atomic snapshots of the (a) SC/MG specimen, (b) CG/MG specimen, and (c) NC/MG specimen.

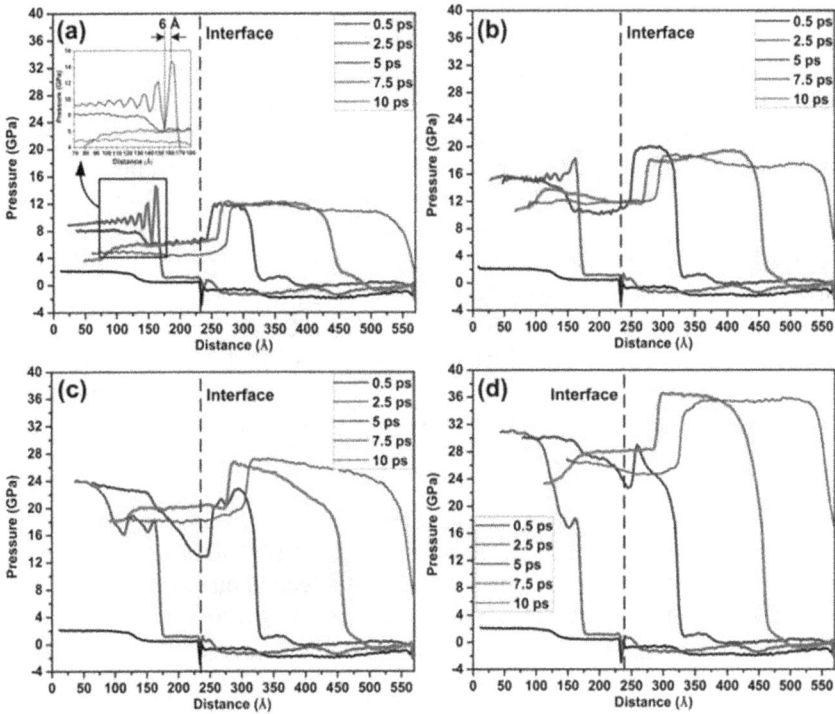

FIGURE 3.23 Pressure profiles of SC/MG specimen for a shock velocity of: (a) 0.5 km/s, (b) 0.8 km/s, (c) 1.1 km/s, and (d) 1.4 km/s. Shock wave propagates from left (crystalline region) to right (amorphous region).

near the amorphous region. The dip in the pressure profile characterized the transition from an elastically compressed region to a plastically compressed region. At a relatively lower piston velocity, the dip was extended to a larger crystalline region as lower velocities aided the elastic compression. As the piston velocity increased, the extended dip in the crystalline region decreased (Figure 3.23(d)). These observations provide valuable insight into the overall deformation behavior of nanolaminates under different shock velocities.

CNA snapshots of NC/MG specimens before and after shock propagation for different piston velocities were analyzed to investigate the structural transformation during the process. The objective was to study the coupling mode between the phase transition and plasticity in each specimen. The study focused on the phase transition in the crystalline region during the process, while the structural evolution in the MG region is discussed later. The length of the final specimen reduced significantly with an increase in piston velocity. Shock compression from the crystalline region at a lower piston velocity resulted in the generation of stacking faults and twin boundaries. In comparison, the stacking faults and twin boundaries were minimal in volume when the shock wave traversed from the amorphous region. Martensitic transformation was observed in the specimen with an increase in piston velocity, and a higher volume fraction of the BCC phase was formed. The martensitic transformation was driven by the nucleation of the BCC phase, which may be through the epitaxial Bain path (also refer to Figure 3.20). The volume of the BCC phase during the shock prop-agation from the crystalline phase was greater than that of the BCC phase during the shock propagation from the amorphous phase. This was due to the generation of rarefaction waves in the former case, which aided in the stabilization of the BCC phase. At a lower piston velocity, the plastic deformation during the shock propaga-tion was controlled through the formation of stacking faults and twin boundaries, but the martensitic transformation was involved. The martensitic transformation was initially triggered near the grain boundaries during the propagation of the shock wave from the crystalline region. Once the shock wave had crossed the interface and the rarefaction wave had released some amount of strain in the crystalline region, the martensitic transformation was reversed, and the mechanism to accommodate the plastic strain shifted back towards the formation of twins and stacking faults. After a similar period of time, the Shockley partial dislocation density showed a gradual increase.

We have also studied the cyclic loading behavior of a nanocrystalline metal-lic alloy system with amorphous intergranular films (AIFs). In recent years, pure nanocrystalline metals have received significant attention due to their potential for enhanced mechanical properties. However, their propensity for brittle fracture poses a significant challenge to their practical use in various applications. In this context, intergranular films have been proposed as an effective means of improving the

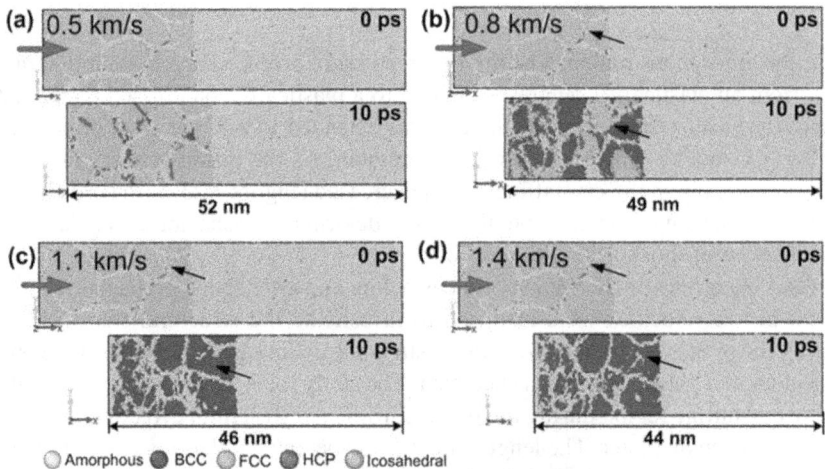

FIGURE 3.24 Atomic snapshot during the structural transformation of NC/MG specimen for different shock velocities. Red arrows show the direction of shock propagation. Black arrows in (b) and (c) show the presence and absence of grain boundaries in the specimen at the initial and final time period, respectively.

toughness and ductility of nanocrystalline metals. The present case study utilized MD simulations to investigate the influence of amorphous intergranular films and their thickness on crack retardation in nanocrystalline Cu under cyclic loading conditions (Figure 3.25).

To further explore the influence of AIF on the mechanical behavior of nanocrystalline Cu under cyclic loading conditions, we conducted simulations on specimens with varying AIF thickness. In particular, we analyzed the mode-I fracture of specimens with edge cracks subjected to increasing strain amplitude fatigue loading over 50 cycles. The cyclic loading conditions applied were similar to those used in previous literature studies on the atomistic behavior of fatigue crack growth in metallic systems. The maximum stress considered was approximately 0.25 of the ultimate tensile strength of the specimen, and the corresponding strain value was set as the maximum tensile strain for the cyclic loading process. The stress–strain plot, presented in Figure 3.26, includes a combination of raw data points obtained from the simulation and polynomial fit curves of the raw data. The polynomial curve fitting of the raw data points was done to obtain a smooth deformation trend during the cyclic loading, with a polynomial fit of order three used in this case.

Our findings showed that the nanocrystalline Cu specimen had the highest tensile strength among all specimens, with and without AIF. The introduction of an AIF with a thickness of 0.5 nm marginally reduced both the tensile strength and plasticity compared to the nanocrystalline Cu specimen. However, specimens with AIF thicknesses of 1 and 2 nm showed an increase in plastic behavior despite a reduction in strength, indicating that the AIF had retarded crack propagation during the 50 cycles (Figure 3.27). Overall, the simulations provide insights into the mechanical behavior of nanocrystalline Cu under cyclic loading conditions and demonstrate the potential for AIF to improve the toughness and ductility of these materials.

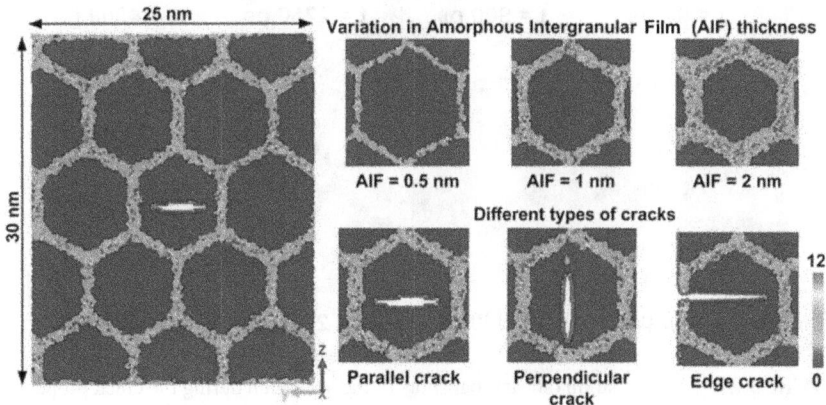

FIGURE 3.25 Atomic snapshot of the NC metallic system with pure interface and AIFs. A pre-designed crack is placed at different positions to see the fatigue failure of the system under cyclic loading.

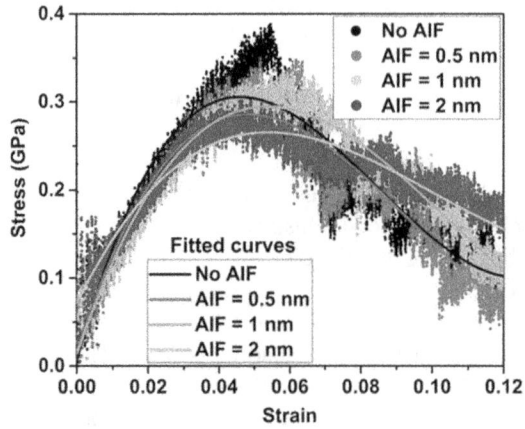

FIGURE 3.26 Stress vs. strain plots for specimens with different AIF thicknesses and edge crack perpendicular to the loading direction (Z-axis) under cyclic loading with a load ratio of 0.5.

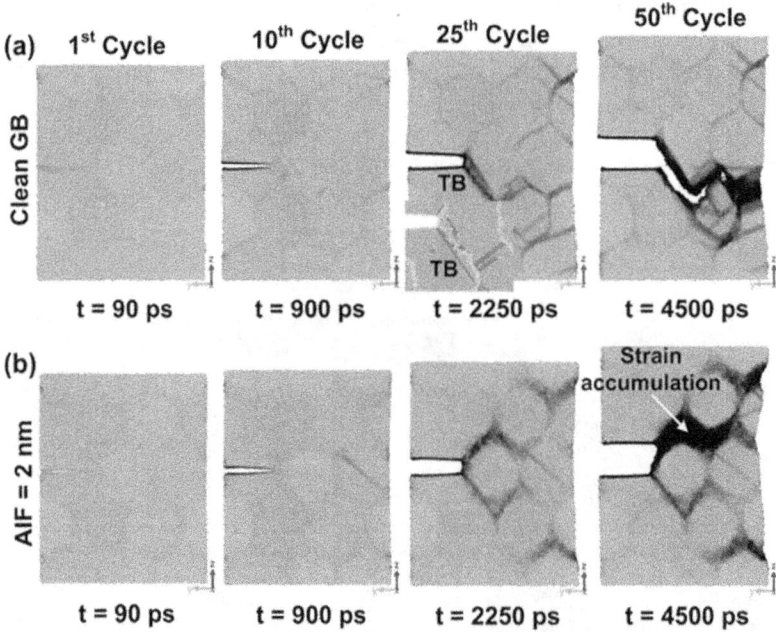

FIGURE 3.27 Atomic strain (shear) snapshots of the specimen during the crack propagation (a) no AIF (nanocrystalline Cu), and (b) AIF with a thickness of 2 nm during a total of 50 cycles.

3.5 EXAMPLE LAMMPS INPUT CODES

Tensile deformation

```
units           metal
echo                    both
atom_style              atomic
dimension               3
boundary                p p p
read_data               read_data_SC_Fe
#for potential file
timestep                0.002
pair_style              eam/fs
pair_coeff              * * MCM2011_eam.fs Fe
thermo                      100
# temperature change
velocity                all create 77 45875 rot yes mom yes dist gaussian
#energy minimization
minimize                1.0e-9 1.0e-6 1000 1000
# this fix is only for equilibration/sample prep
fix             1 all nvt temp 77 77 0.01
run                     1000
unfix                   1
#compute                1 all stress/atom NULL
#compute                2 all reduce sum c_1[1] c_1[2] c_1[3]
#variable               stress equal ((c_2[2])/(3*vol))
variable                tmp equal ly
variable                lo equal ${tmp}
variable                strain equal (ly-v_lo)/v_lo
variable                p1 equal "-pxx/10000"
variable        p2 equal "-pyy/10000"
thermo_style            custom step temp vol etotal pyy lx ly lz v_strain
dump                    1 all custom 5000 Cu_tensile_defo_dump_77K*.lammpstrj
    id type x y z
############ CNA analysis
#dump           3 all custom 5000 stress_peratom* id type x y z c_1[1] c_1[2] c_1[3]
    c_1[4] c_1[5] c_1[6]
log                     Cu_tensile_0.0001_800K.data
# change here the temperature of deformation
fix             3 all nvt temp 77 77 0.01
# temperature change
velocity                all create 77 873847 rot yes mom yes dist gaussian
##Tensile strain rate modification in/ps
fix             4 all deform 1 y erate 0.0001 units box
##Tensile stress strain curve OF DEFORMATION
fix             def all print 200 "${strain} ${p2} ${p1}" file Ni_alloy_stress_
    strain_0.0001_800K_tensile.txt
###### CHANGE BASED ON THE STRAIN RANGE
run                     1000000
```

Creep deformation

```
units           metal
echo            both
atom_style      atomic
dimension       3
boundary        m m m
read_data       read_data_pure_Ni_0_4_Zr
region          fixd block 0 15 INF INF INF INF
region          mobile block 15 585 INF INF INF INF
region          fixd1 block 585 600 INF INF INF INF
region          load block 290 310 INF INF INF INF
group           fixd region fixd
group           fixd1 region fixd1
group           mobile region mobile
group           load region load
timestep        0.002
pair_style      eam/fs
pair_coeff      * * Ni-Zr_Mendelev_2014.eam.fs Ni Zr
thermo          1000
velocity        mobile create 900.0 873847 rot yes mom yes dist gaussian
# Energy minimization
minimize        1.0e-4 1.0e-3 1000 1000
variable        tmp equal ly
variable        lo equal ${tmp}
variable        strain equal (ly-v_lo)/v_lo
variable     time equal step*0.002
variable        yhigh equal yhi
variable        ylow equal ylo
# Equilibration
fix             equi mobile nvt temp 900 900 0.01
run             5000
unfix           equi
# initial velocities
compute             new mobile temp
velocity        mobile create 900 482748 temp new
fix             1 mobile nvt temp 900.0 900.0 0.05
fix             11 fixd rigid single
fix             2 fixd setforce 0.0 0.0 0.0
fix             31 fixd1 rigid single
fix             3 fixd1 setforce 0.0 0.0 0.0
thermo_style    custom step temp xlo xhi ylo yhi zlo zhi etotal press pyy vol
dump            1 all custom 5000 Ni_bicrystal_NC_creep_dump@900K*.
    lammpstrj id type x y z
log             logNi_bicrystal_NC_creep_dump@510K.data
fix             4 mobile nve
#compute        csym all centro/atom fcc
#compute        peratom all pe/atom
############# CNA analysis (id replace by mass for linex)
```

```
#dump           CSP all cfg 2000 dump.Ni-Zr_NC_creep_dump@900K*.cfg
   mass type xs ys zs c_csym c_peratom fx fy fz
fix             def mobile print 500 "${time} ${ylow} ${yhigh} ${strain}" file
   Ni_bicrystal_creep_time_strain_curve_data.txt
fix             7 load addforce 0 0.006 0
run             4000000
```

4 Molecular Dynamics Simulation of Metallic Glass

4.1 INTRODUCTION TO METALLIC GLASSES

Various solid-state materials, featuring different bond types such as ionic, covalent, van der Waals, hydrogen, and metallic, can be transformed into amorphous solid forms using various techniques. Among these amorphous materials, metallic glasses are relatively new. The first metallic glass, $Au_{75}Si_{25}$, was developed in 1960 by Duwez at Caltech, USA, through the rapid quenching of metallic liquids at very high cooling rates of 10^5–10^6 K/s [Klement et al., 1960]. This method bypassed the usual nucleation and growth of crystalline phases, resulting in a frozen liquid configuration. Since then, research on metallic glasses has gained momentum, especially in the early 1970s and 1980s, when continuous casting processes were developed for the commercial production of metallic glass ribbons, lines, and sheets. However, due to the high cooling rates required, the geometry of amorphous alloys was limited to thin sheets and lines, which had limited applications. Turnbull and his colleagues played a significant role in the development of metallic glasses by demonstrating the similarities between metallic glasses and other non-metallic glasses such as silicates, ceramic glasses, and polymers [Turnbull, 1961]. They also showed that a glass transition observed in conventional glass-forming melts could also be observed in rapidly quenched metallic glasses. The glass transition occurred at a well-defined temperature that varied only slightly as the heating rate was changed. Turnbull's reduced glass transition temperature ratio, $T_{rg} = T_g/T_m$, which determines the glass-forming ability (GFA) of an alloy, has been a valuable criterion for predicting the GFA of any liquid. This criterion has played a key role in the development of various metallic glasses, including bulk metallic glasses (BMGs).

The typical strength and elastic limit of a material depend on its type, composition, and processing method. Generally, metals and alloys have high strength and elastic limits, while non-metals, such as ceramics and polymers, have lower strength and elastic limits (Figure 4.1). However, metallic glasses are unique in this regard. Unlike crystalline metals and alloys, which exhibit plastic deformation through slip and dislocation mechanisms, metallic glasses exhibit homogeneous deformation through the collective movement of atoms. This unique mechanism results in high strength and elastic limits for metallic glasses, comparable to those of crystalline metals and alloys. The strength of metallic glasses is typically in the range of 1–4 GPa, depending on the composition and processing method. The elastic limit of metallic glasses is also high, typically in the range of 1–2 GPa. These values are

DOI: 10.1201/9781003323495-4

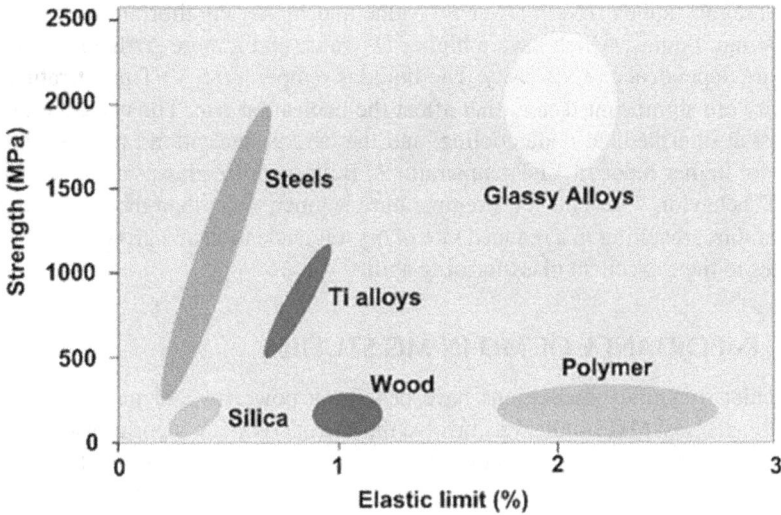

FIGURE 4.1 Typical strengths and elastic limits for different materials. The metallic glasses show that they are unique.

comparable to or even higher than those of some high-strength steels and titanium alloys. Furthermore, metallic glasses have excellent fracture toughness and fatigue resistance, which make them attractive for a wide range of engineering applications, such as aerospace, biomedical, and consumer electronics industries.

To achieve glass formation in metallic liquids, it is essential to understand the thermodynamic and kinetic aspects of crystallization. One of the key parameters for assessing the glass-forming ability of metallic liquids is the reduced glass transition (T_{rg}), which represents the ratio of the glass transition temperature (T_g) to the liquidus temperature (T_{liq}). When a liquid is cooled below its melting point, the free energy difference between the liquid and a crystal acts as a driving force for crystal nucleation. However, the formation of the liquid–crystal interface generates positive interfacial energy, which hinders nucleation. This creates an energy barrier that local composition fluctuations must overcome to form a nucleus. The growth of a nucleus requires atomic rearrangement within the liquid, and the rate of atomic transport is characterized by the atomic diffusivity or viscosity.

The process of crystal nucleation in metallic liquids is influenced by both thermodynamic and kinetic factors. The thermodynamic factor considers the likelihood of overcoming the nucleation barrier, while the kinetic factor is determined by atomic diffusion or viscosity. A formula that considers both factors determines the nucleation rate, which takes into account the Gibbs free energy difference, the interfacial energy between the liquid and nuclei, and the nucleation barrier for forming a spherical nucleus. Viscosity plays a crucial role in glass formation, which is often described by the Vogel–Fulcher–Tamman (VFT) relation modified by the fragility parameter

(D^*). Fragile liquids have a lower D^* value and higher equilibrium melt viscosity than strong liquids, which have a higher D^* value and a more Arrhenius-like temperature dependence of viscosity. The liquidus temperature, VFT temperature, and fragility are significant factors that affect the nucleation rate. The nucleation rate is highest at intermediate undercooling, and the crystal nucleation rate is suppressed more for higher reduced VFT temperatures. Bulk metallic glasses exhibit a "strong liquid" behavior, where the VFT temperature is much lower than the glass transition temperature, resulting in a reduced rate of crystal nucleation and growth, which contributes to their excellent glass-forming ability.

4.2   IMPORTANCE OF MD IN MG STUDIES

Molecular dynamics simulations have become a powerful tool in the analysis of metallic glasses. MD simulations involve the use of computer algorithms to simulate the motion of atoms and molecules within a material. By providing a detailed understanding of the behavior of metallic glasses at the atomic level, MD simulations can help to predict and explain their unique glassy properties. In this section, we will discuss the importance of MD simulations in metallic glass analysis and review some of the recent literature on this topic.

One of the key advantages of MD simulations is their ability to capture the complex and highly non-equilibrium nature of metallic glass formation. Traditional experimental techniques, such as X-ray diffraction or differential scanning calorimetry, provide important information about the thermodynamics of metallic glass formation but are limited in their ability to capture the kinetic processes involved. MD simulations, on the other hand, can model the evolution of nanoclusters in metallic glass from the melt and provide detailed insight into the underlying atomic processes. For example, recent MD simulations have shed light on the role of local atomic packing in controlling metallic glass formation. By analyzing the Voronoi tessellation of the atomic structure, researchers have shown that the formation of a high density of icosahedral structures is crucial for promoting metallic glass formation [Wang et al., 2016]. Other studies have highlighted the importance of atomic diffusion and shear banding in determining the final microstructure of metallic glasses [Sheng et al., 2022; Zhong et al., 2016]. By capturing these complex processes at the atomic level, MD simulations can provide a more complete understanding of metallic glass formation than is possible through traditional experimental techniques. MD simulations can also help to guide experimental design by predicting the properties of metallic glasses that are difficult or impossible to measure directly. For example, MD simulations have been used to predict the mechanical properties of metallic glasses, such as their strength and ductility, and to investigate the effect of alloy composition on these properties [Amigo et al., 2023]. Similarly, MD simulations can predict the glass transition temperature and the fragility of metallic glasses, which are difficult to measure experimentally [Wu et al., 2021; Senko, 2007]. Moreover, MD simulations can also be used to design new metallic glass alloys with improved properties. By simulating the formation and properties of hypothetical metallic glasses, researchers can identify new compositions and processing routes that are likely to exhibit improved glass-forming ability or other desirable properties. For example, recent

MD simulations have identified new metallic glass alloys with ultra-high strength and toughness [Kruzic, 2016]. In the subsequent sections, we will elaborate on the design of metallic glasses using MD simulation and analyze the effect of operating parameters and corresponding evolution in the properties to understand the underlying physics behind such behavior.

4.3 DESIGNING METALLIC GLASSES USING MD SIMULATION IN LAMMPS

In this section, we will discuss the preparation of MG specimens and the parameters that could affect metallic glass formation. We will consider the Ni-Nb alloy system, which is a good glass-forming alloy, as a case study. The objective is to explore how the glass-forming ability and plasticity of Ni-Nb metallic glass are affected by the addition of Nb through the use of molecular dynamics simulation employing embedded atom method interatomic potential. To prepare the specimens, a fast cooling process was employed from the liquid phase for Ni-Nb metallic glass with varying at. % of Nb (i.e., 15, 20, 25, 30, 38, 40.5, 45, and 50 at. %), as shown in Figure 4.2. To create the metallic glass specimens, they were first heated to a temperature of 3000 K and held at this temperature for 20 ps, resulting in complete melting and the absence of any crystalline phase. Next, the specimens were rapidly cooled at a rate of 10^{12} K s^{-1} to form an amorphous metallic glass structure. Both the heating and cooling processes were carried out under an NPT ensemble (N is number of atoms, P is pressure, and T is temperature) with periodic boundary conditions, and a time step of 0.002 ps was used for the simulation.

The glass transition temperature (T_g) of all Ni-Nb metallic glass specimens was determined using a volume versus temperature plot during the rapid cooling process, as depicted in Figure 4.3. The intersection point of the linear fit of volume at different temperatures was used to determine T_g for each specimen. The high cooling rate

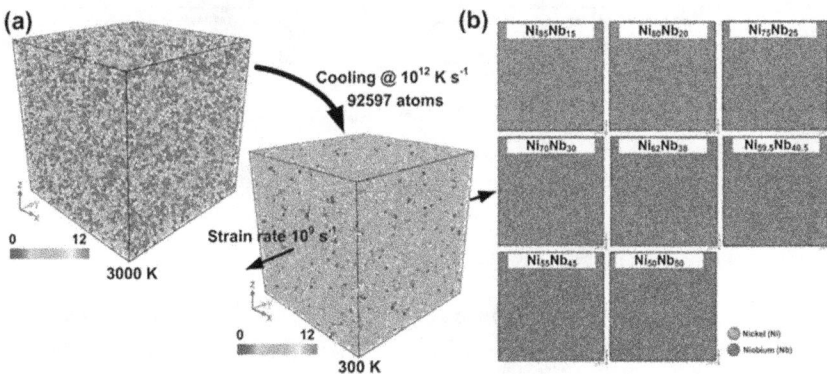

FIGURE 4.2 (a) Atomic snapshots of the cooling process of Ni-Nb metallic glass with varying at. % of Nb, along with the representation of the tensile loading direction on the Y-axis. (b) Atomic snapshots of the cross-section of all the Ni-Nb metallic glass specimens with varying at. % of Nb.

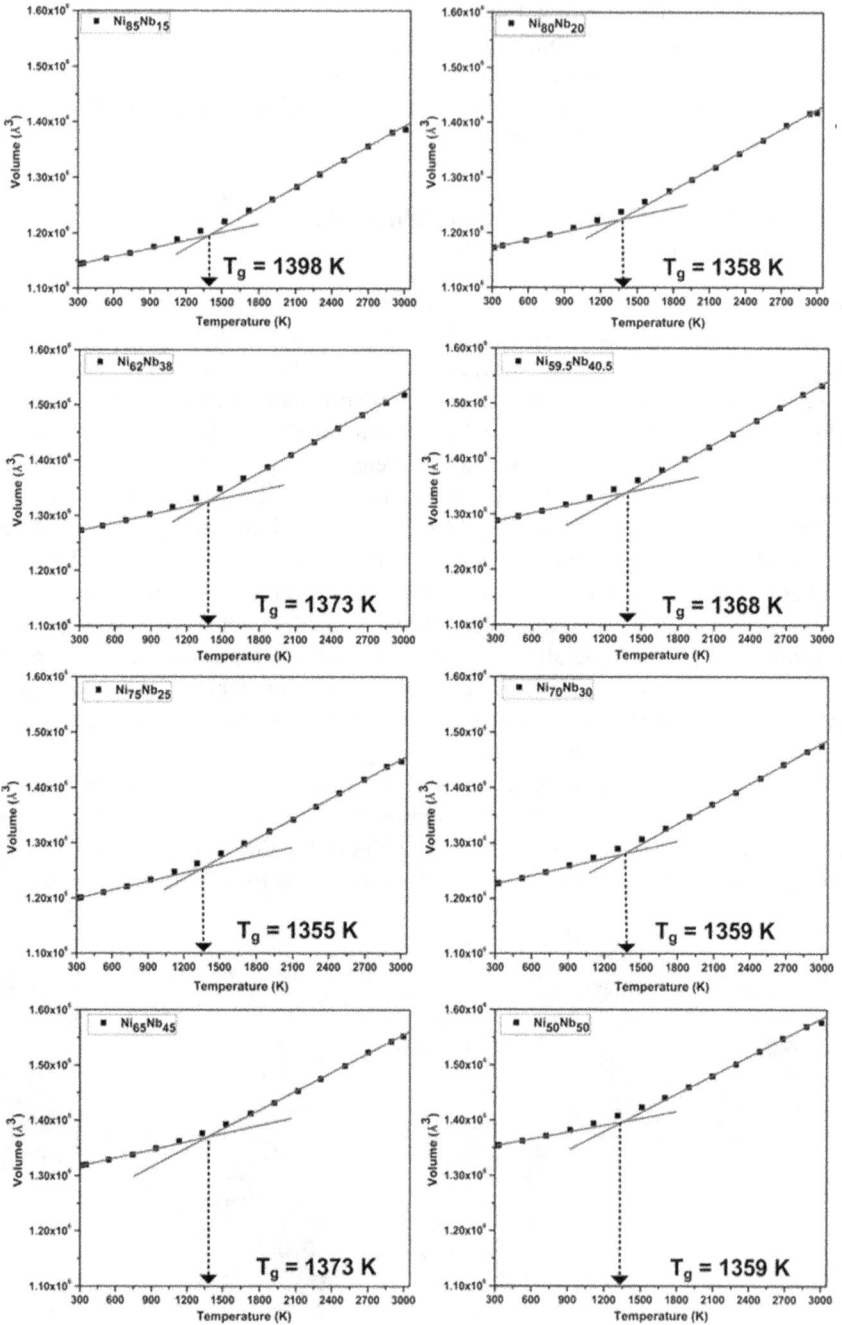

FIGURE 4.3 Volume vs. atomic percent of Nb plots for all Ni-Nb metallic glass specimens. The glass transition temperature (T_g) is indicated at the point of sudden volume change

(10^{12} K s^{-1}) utilized in this study resulted in the formation of an amorphous structure with excess free volume, leading to a sudden change in volume that indicates T_g. However, the molecular dynamics simulation values obtained were slightly higher than experimental measurements due to the absence of enough time for structural relaxation. An increase in the atomic percentage of Nb led to an increase in the volume of the Ni-Nb metallic glasses due to the larger atomic size of Nb in comparison to Ni.

The liquidus temperature (T_{liq}) for each specimen was determined using a slow cooling rate of 0.5 K s^{-1} up to 2100 K while monitoring the volume change in the crystalline Ni-Nb systems. The liquidus temperature values were then compared with the Ni-Nb binary phase diagram obtained from the CALPHAD method using Thermo-Calc software (Figure 4.4). The obtained MD simulation temperatures showed good agreement with liquidus temperature values obtained from the bulk phase diagram, except for the Ni$_{85}$Nb$_{15}$ and Ni$_{80}$Nb$_{20}$ systems. The deviation was attributed to the low concentration of Nb atoms in nanostructured alloys that cannot distort the lattice at lower temperatures or form an amorphous structure. However, as the concentration of Nb atoms increased, the deviation of liquidus temperature from the bulk phase diagram was reduced. This approach has been widely used to determine the phase equilibria and thermodynamic properties of metallic systems as well as the glass-forming ability of metallic glasses.

The metallic glass specimens' yield strength and ultimate tensile stress were evaluated using tensile loading. The tensile stress–strain plots were used to determine the yield stress and ultimate tensile stress for each specimen using the 0.2% strain-offset method and the highest stress value, respectively (Figure 4.5). The relationship between the Nb content and tensile strength, ultimate tensile stress, and plasticity was then analyzed (Figures 4.6(a–c)). The results revealed that the tensile strength and ultimate tensile stress initially decreased with increasing Nb content up to 30 at. %

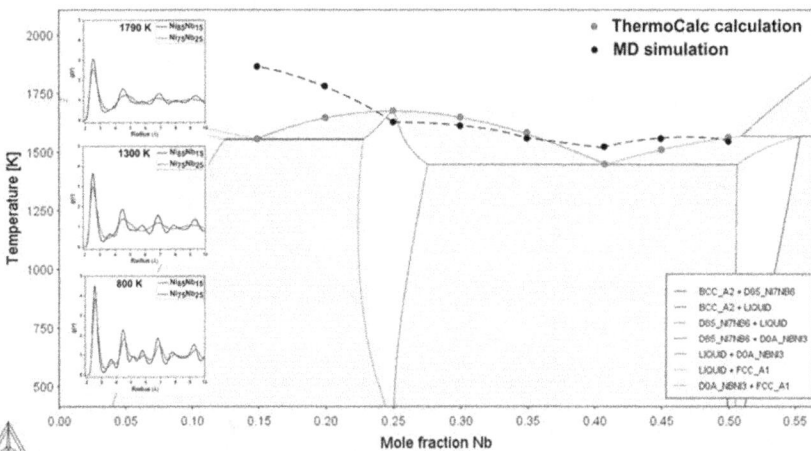

FIGURE 4.4 Comparison of liquidus temperature for Ni-Nb metallic glass specimens obtained from Thermo-Calc and molecular dynamics simulation. The inset shows a comparative radial distribution function plot of Ni$_{85}$Nb$_{15}$ and Ni$_{75}$Nb$_{25}$ specimens.

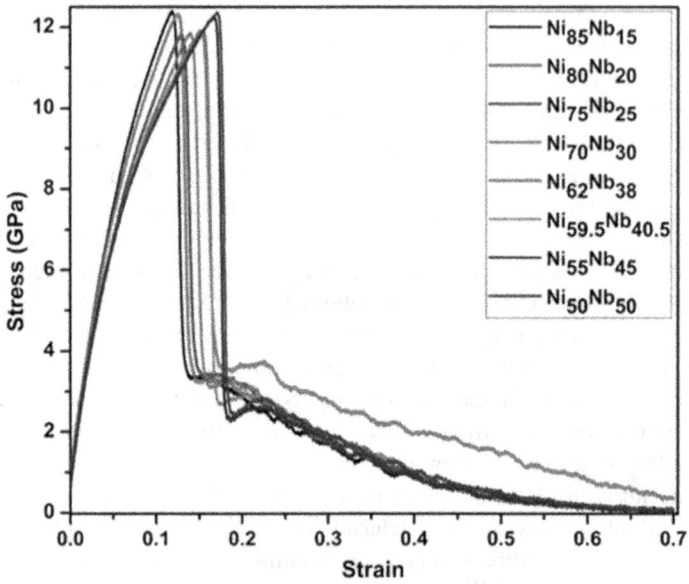

FIGURE 4.5   Stress vs. strain plots of all Ni-Nb metallic glass specimens.

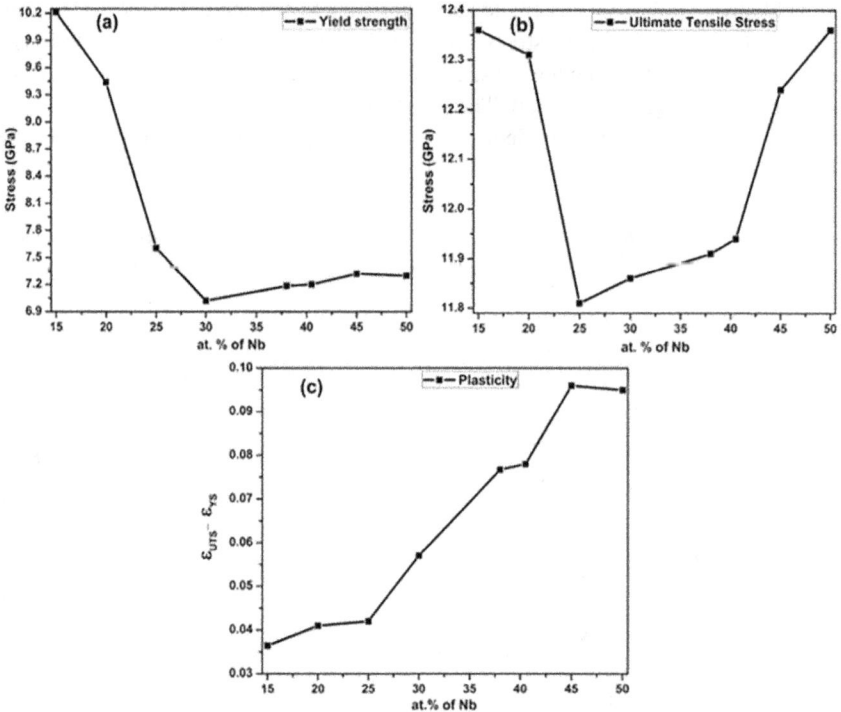

FIGURE 4.6   Plot of: (a) variation of yield strength with increasing at. % of Nb, (b) variation of ultimate tensile stress with increasing at. % of Nb, and (c) variation of plasticity (difference between ultimate tensile strain (ε_{UTS}) and yield strain (ε_{YS})) with increasing at. % of Nb.

and then slightly increased beyond this concentration. The number of icosahedral clusters with index <0 0 12 0> in the specimens explained these trends, as it influences the metallic system's tensile strength. A higher number of Ni-centered icosahedral clusters at lower Nb content led to higher yield strength and ultimate tensile strength. However, an increase in the number of icosahedral clusters reduced the metallic glass specimen's ductility. Plasticity, calculated as the difference between the yield strain and the strain at maximum stress, increased with an increase in the Nb content. The voids generated in the specimens at different strain values during tensile deformation were analyzed using surface mesh analysis, revealing that larger voids were formed in the $Ni_{85}Nb_{15}$ and $Ni_{80}Nb_{20}$ specimens at 13% strain, while smaller voids were observed in the $Ni_{75}Nb_{25}$ and $Ni_{70}Nb_{30}$ specimens. The specimens with Nb content > 30 at. % showed voids at higher strain values (15% strain). The $Ni_{55}Nb_{45}$ and $Ni_{50}Nb_{50}$ specimens exhibited the highest plasticity, as voids were not observed even at 15% strain, indicating that an increase in Nb content enhanced the metallic glass specimen's plasticity.

4.4 VORONOI TESSELLATION METHOD

The Voronoi tessellation method is a popular tool for analyzing the structure of metallic glasses. This method is based on the idea of partitioning space into Voronoi cells around each atom in the metallic glass system. The Voronoi cell around a given atom is defined as the set of points in space that are closer to this atom than to any other atom in the system. Using this method, it is possible to obtain a detailed description of the atomic packing structure in metallic glasses. In particular, the Voronoi analysis provides information about the number and size of different types of structural motifs, such as icosahedral clusters and polyhedral clusters. These motifs are known to play an important role in determining the mechanical properties of metallic glasses. Moreover, the Voronoi tessellation method can be used to study the evolution of the atomic structure during deformation and relaxation processes. For example, it is possible to track the changes in the Voronoi cell volumes and shapes as the metallic glass is subjected to external stresses or thermal annealing. In a Voronoi diagram, each data point is assigned to the nearest centroid or center of a cluster, creating a polygon around it, and hence the group of atom clusters can be identified based on the sharing between them. For instance, in a face-sharing Voronoi cluster, adjacent polygons share a common edge. This means that the clusters have a common boundary, which may be useful in certain applications where adjacent clusters may have similar characteristics. On the other hand, in an edge-sharing Voronoi cluster, the polygons do not necessarily share a common edge but rather share points along their edges. This may be useful in cases where there are sharp boundaries between clusters or where the data points are distributed in a more irregular manner. These clusters are identified by a set of numbers $<n_3, n_4, n_5, n_6>$, which represents the number of polygons (n) that are three-faced (triangular), four-faced, five-faced (pentagon-shaped), and six-faced (hexagonal-shaped). A value of n that is 1 or 2 is not used, and it is left up to the readers to do a little brainstorming.

In addition to face-sharing and edge-sharing Voronoi clusters, there is also a type of Voronoi clustering known as volume-sharing Voronoi clusters. This approach partitions the data points into regions based on the volume of space surrounding each

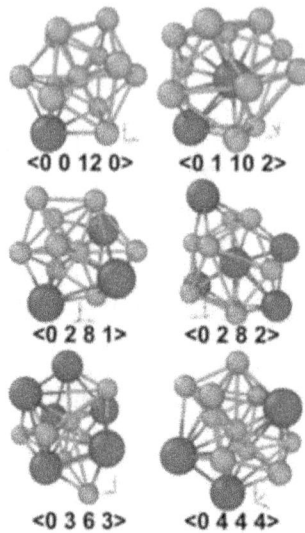

$\langle 0\ 0\ 12\ 0 \rangle$ $\langle 0\ 1\ 10\ 2 \rangle$

$\langle 0\ 2\ 8\ 1 \rangle$ $\langle 0\ 2\ 8\ 2 \rangle$

$\langle 0\ 3\ 6\ 3 \rangle$ $\langle 0\ 4\ 4\ 4 \rangle$

FIGURE 4.7 Atomic configurations of various icosahedral-like clusters in metallic glass specimens.

point rather than just the area on a two-dimensional plane. MD simulation techniques are crucial and insightful in determining and studying these clusters at different temperatures, and they can critically analyze the evolution of these clusters to give a better understanding of metallic glass formation. We present a case study that discusses Voronoi tessellation in metallic glass in detail [Reddy and Pal, 2019].

In this study, we investigated the impact of composition on the formation of various Voronoi clusters in a Zr-Nb metallic alloy system. Prior literature has shown that the $Zr_{50}Nb_{50}$ alloy has the highest glass-forming ability [Reddy and Pal, 2019]. To gain a better understanding of the distribution and connectivity of icosahedral and distorted icosahedral clusters in the $Zr_{50}Nb_{50}$ specimen at 300K, we present Figure 4.8. Previous research has established that clusters can be connected through vertex sharing, edge sharing, face sharing, and volume sharing [Soklaski et al., 2013; Trady et al., 2017]. The figure illustrates the vertex and face sharing of the Nb-centered icosahedral cluster in the $Zr_{50}Nb_{50}$ specimen (Figures 4.8(a) and (b), respectively). In addition, Figures 4.8(c) and (d) demonstrate the connectivity of two interpenetrating icosahedral clusters through edge sharing and vertex sharing, respectively. These interpenetrating icosahedral clusters contribute to the dense atomic packing of metallic glasses with good GFA [Hirata et al., 2013]. In the $Zr_{50}Nb_{50}$ specimen, these clusters are connected through edge and vertex sharing, which enhances their connectivity. Furthermore, the atomic snapshot obtained from the MD simulation indicates that the icosahedral and distorted icosahedral clusters are connected through face and volume sharing. This shows the presence of a Zr-centered volume sharing $\langle 0\ 2\ 8\ 5 \rangle$ cluster, indicating that distorted icosahedral clusters can also form closed packing structures in metallic glasses. Finally, Figures 4.8(e), (g), and (h) reveal the presence of both Nb-centered and Zr-centered clusters in the volume-sharing

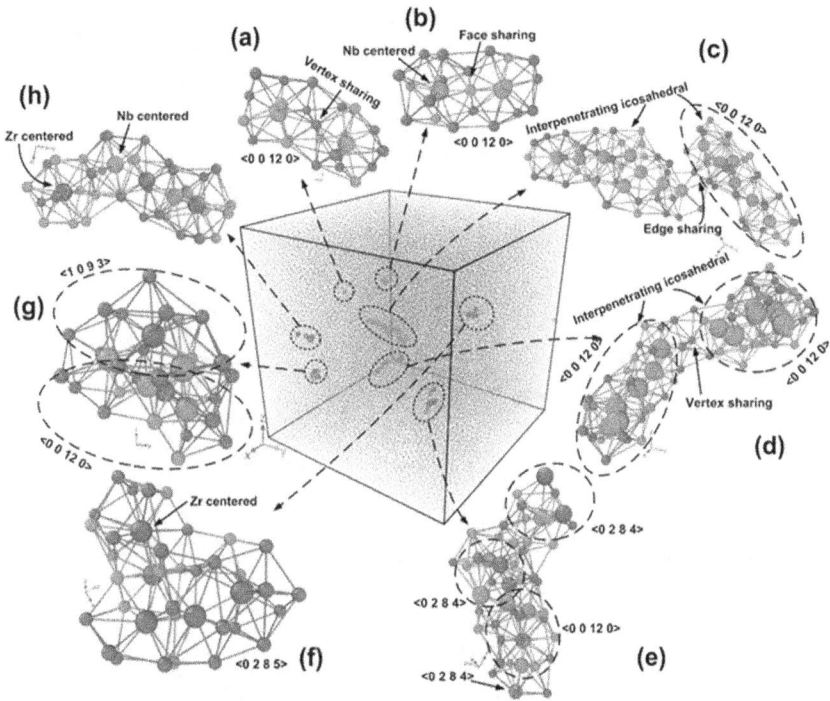

FIGURE 4.8 Illustration of different types of clusters and their connectivity through vertex, edge, face, and volume sharing for a $Zr_{50}Nb_{50}$ metallic glass specimen.

icosahedral and distorted icosahedral clusters, demonstrating structural uniformity in the metallic glass specimen.

In addition to examining the distribution and connectivity of clusters in a $Zr_{50}Nb_{50}$ metallic glass sample after cooling it to 300 K, it is important to consider the formation of intermediate clusters during the cooling process. These intermediate clusters, such as $\langle 0\,2\,8\,4\rangle$ and $\langle 1\,1\,9\,0\rangle$, transform gradually into the $\langle 0\,0\,12\,0\rangle$ cluster, and this transformation can be traced using molecular dynamics simulations. Figure 4.9 shows snapshots of the atomic structures of Voronoi clusters during the cooling process and their transformations. For example, in Figure 4.9(a), the $\langle 0\,2\,8\,4\rangle$ cluster transforms into the $\langle 0\,0\,12\,0\rangle$ cluster through the intermediate $\langle 1\,0\,9\,3\rangle$ cluster. During this transformation, the $\langle 0\,2\,8\,4\rangle$ cluster rejects an atom and changes into the $\langle 1\,0\,9\,3\rangle$ cluster, which then transforms into the $\langle 0\,0\,12\,0\rangle$ cluster by removing a triangular-faced edge. This transformation mechanism has also been observed in previous studies. Similarly, Figure 4.9(b) shows that adding an extra Zr atom transforms the $\langle 1\,1\,9\,0\rangle$ cluster into the $\langle 0\,2\,8\,2\rangle$ cluster, which then restructures itself to form the $\langle 0\,0\,12\,0\rangle$ cluster. Additionally, removing two Zr atoms from the $\langle 0\,2\,8\,4\rangle$ cluster transforms it into the $\langle 0\,2\,8\,2\rangle$ cluster, which also restructures itself to form the $\langle 0\,0\,12\,0\rangle$ cluster, as shown in Figure 4.9(c). It is possible for a particular cluster to follow different paths during the structural transformation process, as illustrated in Figure 4.9(d), which displays

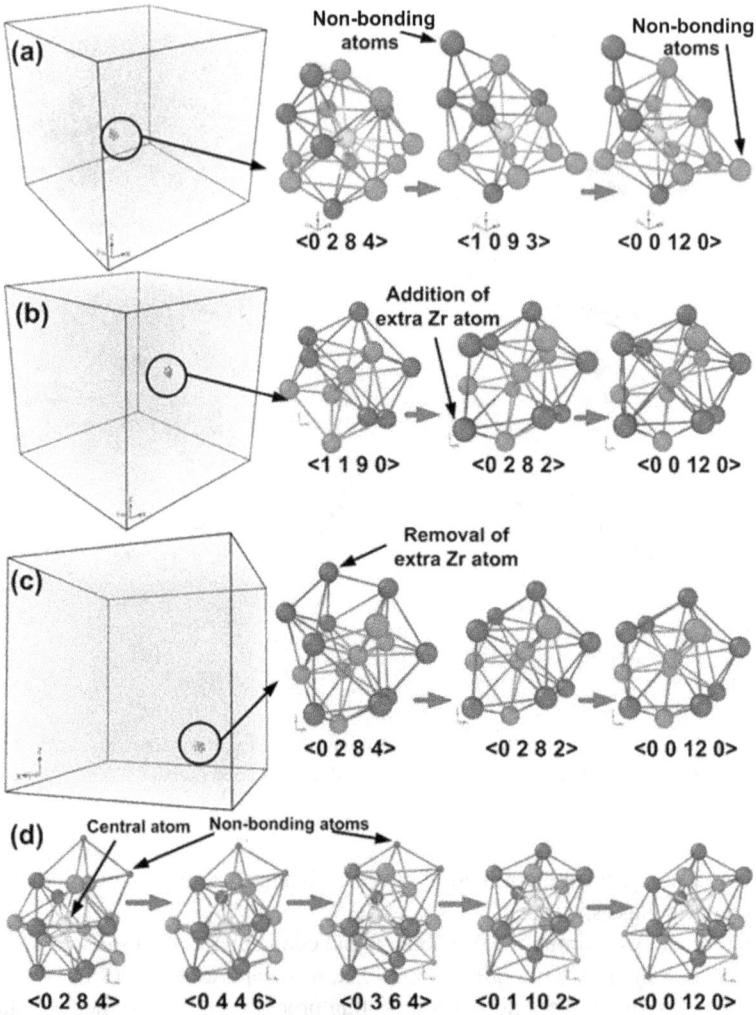

FIGURE 4.9 Atomic snapshots illustrating the structural transformation of various Voronoi polyhedra during the cooling process of a $Zr_{50}Nb_{50}$ metallic glass specimen. The figure shows the transformation pathways for the $\langle 0\ 2\ 8\ 4 \rangle$, $\langle 1\ 1\ 9\ 0 \rangle$, and $\langle 0\ 2\ 8\ 2 \rangle$ clusters into the $\langle 0\ 0\ 12\ 0 \rangle$ cluster, as well as an alternative pathway for the transformation of the $\langle 0\ 2\ 8\ 4 \rangle$ cluster through intermediate $\langle 0\ 4\ 4\ 6 \rangle$ and $\langle 0\ 3\ 6\ 4 \rangle$ clusters.

atomic snapshots of the $\langle 0\ 2\ 8\ 4 \rangle$ cluster and its transformation into the $\langle 0\ 0\ 12\ 0 \rangle$ cluster. In this case, intermediate $\langle 0\ 4\ 4\ 6 \rangle$ and $\langle 0\ 3\ 6\ 4 \rangle$ clusters are formed before the formation of the $\langle 0\ 1\ 10\ 2 \rangle$ cluster, which then transforms into the $\langle 0\ 0\ 12\ 0 \rangle$ cluster by removing an atom. Finally, after analyzing the structural transformation of the Zr-Nb metallic glass system in detail, a possible transformation route is presented in Figure 4.10, which illustrates the formation of various metastable clusters during the glass cooling process. This schematic diagram shows the possible transformation of five

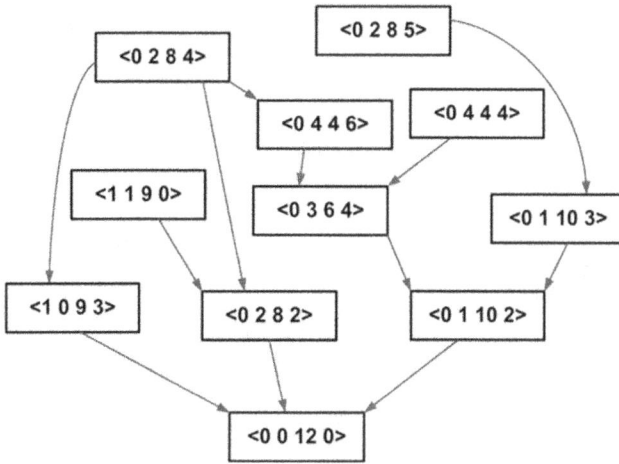

FIGURE 4.10 Pathway of structural transition from distorted icosahedral clusters to icosahedral clusters observed during the cooling process of a Zr-Nb specimen.

major Voronoi polyhedra, $\langle 0\ 2\ 8\ 2 \rangle$, $\langle 0\ 2\ 8\ 4 \rangle$, $\langle 0\ 2\ 8\ 5 \rangle$, $\langle 1\ 0\ 9\ 3 \rangle$, and $\langle 0\ 1\ 10\ 2 \rangle$, into $\langle 0\ 0\ 12\ 0 \rangle$ polyhedra during the cooling process. The schematic diagram reveals that the formation of icosahedral clusters, specifically the $\langle 0\ 0\ 12\ 0 \rangle$ cluster, is primarily due to the transformation of the $\langle 1\ 0\ 9\ 3 \rangle$, $\langle 0\ 1\ 10\ 2 \rangle$, and $\langle 0\ 2\ 8\ 2 \rangle$ clusters. These analyses can only be possible using atomistic simulation, as such intricate detailing of alteration in clusters with high temperatures and a rapid cooling process is almost impossible with experimental procedures at the present time.

4.5 EVALUATION OF PHYSICAL PROPERTIES OF MG

In this section, we discuss the evaluation of liquidus/melting temperatures, viscosity, and fragility calculations for Nb-Zr metallic glasses using MD simulations. It is known that while heating a metallic specimen, the volume changes due to expansion, and upon cooling, it decreases. The change in volume also gives insight into the change in the phase from solid to liquid, and from the analysis, we can determine the liquidus temperatures. The alteration in volume concerning temperature during the heating process of Zr-Nb specimens is presented in Figures 4.11(a–c). In all instances, a sharp decline in volume is noticeable, suggesting the melting of the metallic specimens. The plot is utilized to identify the liquidus temperature (T_{liq}), which is then compared with the phase diagram obtained from ThermoCalc, as illustrated in Figure 4.11(d). ThermoCalc is a thermodynamic database software package based on the CALPHAD method, widely recognized for determining phase diagrams with accuracy [Shi et al., 2005]. The CALPHAD method is a reliable approach for exploring the thermodynamic properties and phase transformation of metallic systems [Antonov et al., 2015]. The obtained results show that the liquidus temperatures of Zr-Nb specimens from both the MD simulation and ThermoCalc calculations are

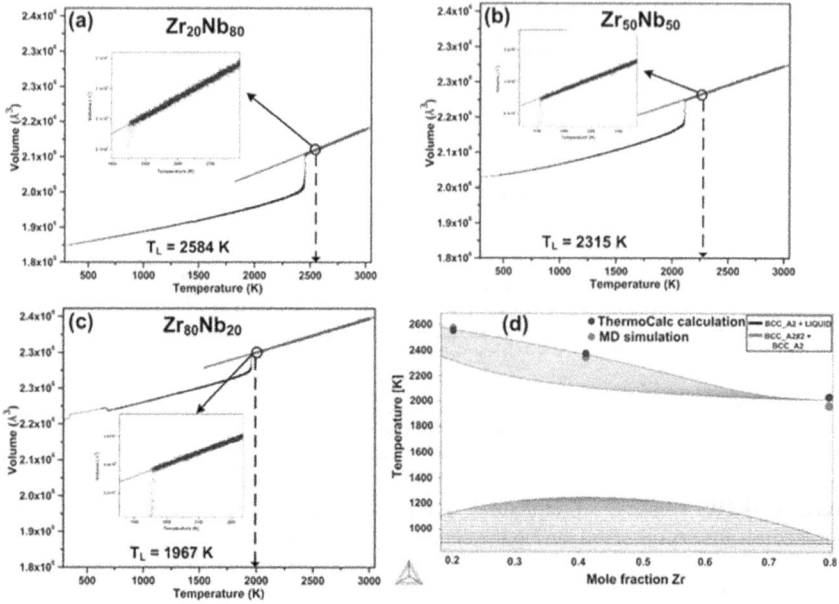

FIGURE 4.11 Determination of liquidus temperature (T_{liq}) for Zr-Nb specimens. (a) Plots of volume vs. temperature for $Zr_{20}Nb_{80}$, $Zr_{50}Nb_{50}$, and $Zr_{80}Nb_{20}$ specimens. (b) Comparison of T_L obtained from MD simulation with that calculated using ThermoCalc.

comparable, indicating the reliability of the angular-dependent potential (ADP) utilized in this study for examining heating and cooling processes.

In this section, we have determined the liquid fragility of the Zr-Nb systems by analyzing the viscosity against the reduced transition temperature (T_g/T) using a semi-log plot. To calculate the viscosity, we utilized the Green–Kubo (GK) method, which involves computing the average of the auto-correlation of stress/pressure tensor at various temperatures during the cooling process. This enabled us to determine the viscosity of the material, and these calculations were carried out for all three specimens in the study. The results were plotted to determine the liquid fragility of Zr-Nb systems. Figure 4.12 displays Angell's plot, which is a logarithmic plot of viscosity versus reduced transition temperature. The slope of the curves at $T_g/T = 1$ represents the fragility index (*m*), where a higher value of *m* indicates a fragile liquid with low glass-forming ability. In this study, we utilized a second-order polynomial fitting function to obtain the curve from the viscosity data points. Our results show that the logarithmic viscosity curve of the $Zr_{20}Nb_{80}$ specimen deviates from Arrhenius behavior, whereas the viscosity change in the $Zr_{50}Nb_{50}$ specimen is almost linear. As the Nb content decreases further, the deviation from Arrhenius behavior increases. The fragility index values of the Zr-Nb specimens were also calculated, and we found that the $Zr_{50}Nb_{50}$ specimen had the highest glass-forming ability, with

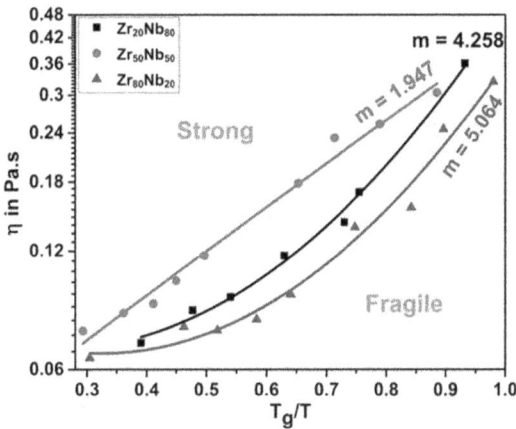

FIGURE 4.12 A semi-log plot of viscosity as a function of temperature to determine the liquid fragility of Zr-Nb metallic specimens.

the lowest value of $m = 2$. In contrast, the $Zr_{20}Nb_{80}$ specimen can be classified as a fragile liquid with low glass-forming ability, since it has a higher value of $m = 4$. The $Zr_{80}Nb_{20}$ specimen shows an intermediate glass-forming ability. These results are consistent with the literature, where a lower value of the fragility index (m) signifies a higher glass-forming ability.

MD simulation is also a useful tool for analyzing physical properties such as the glass transition temperature of materials under different extreme external conditions. A study on an Al-Sm alloy system was conducted to investigate its glass transition behavior under varying hydrostatic pressures (Figure 4.13) [Mishra and Pal, 2018]. The alloy was first melted at a high temperature of 1750 K, well above its melting temperature, to obtain a molten alloy. The cooling process was then performed at different hydrostatic pressures, and the variation of volume as a function of temperature was plotted to determine the glass transition temperature (T_g). T_g was obtained by finding the intersection of the linearly fitted lines of the volume–temperature variation data. The T_g value obtained for the $Al_{90}Sm_{10}$ alloy under 0 GPa pressure was found to be 645 K, which was consistent with previous studies. However, slight variations in T_g could be attributed to differences in simulating parameters such as temperature, cooling rate, and holding time. Interestingly, the T_g value reported in this study was closer to the experimental value for an $Al_{92}Sm_8$ alloy. The observed increase in T_g with increasing hydrostatic pressure was consistent with similar findings in other metallic glass-forming liquids. The phenomenon could be explained by improved local ordering and a reduction in the average atomic dynamics under the influence of hydrostatic pressure. It is important to note that discrepancies between experimental and simulation-based studies may arise from factors such as very high cooling rates or overestimation of the order of Al-Sm liquid alloys during MD study using the VASP package for potential development.

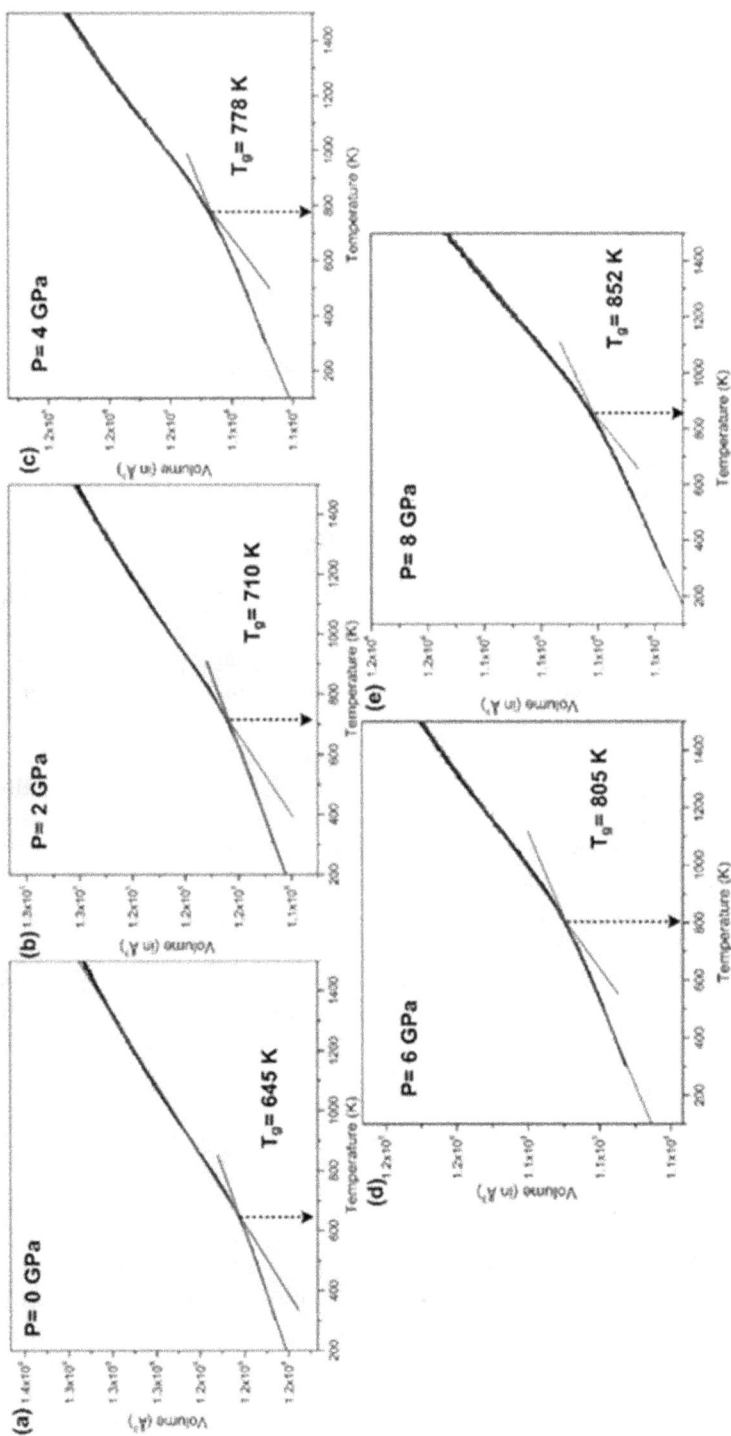

FIGURE 4.13 Volume *vs* temperature plots for the $Al_{90}Sm_{10}$ metallic glass specimen at different hydrostatic pressures: (a) P = 0 GPa, (b) P = 2 GPa, (c) P = 4 GPa, (d) P = 6 GPa, and (e) P = 8 GPa. The glass transition temperature is indicated by the point of change in volume.

4.6   EXAMPLE LAMMPS INPUT CODES

Metallic glass preparation

```
units           metal
echo                    both
atom_style              atomic
dimension               3
boundary                p p p
read_data               read_data_SC_Ti
region                  Ti block INF INF INF INF INF INF units box
set                     region Ti type/fraction 2 0.66666666 1239393
timestep                0.002
pair_style              meam/spline
pair_coeff              * * TiO.meam.spline Ti O
# Energy Minimization energy force
minimize                1.0e-12 1.0e-12 1000 1000
thermo                  100
thermo_style            custom step temp vol press pe ke etotal
#compute        csym all centro/atom fcc
#compute        peratom all pe/atom
dump                    1 all custom 10000 TiO_dump_new*.lammpstrj id type x y z
#CNA analysis
#dump                   2 all cfg 50000 dump.Cu_Zr_dump_new*.cfg mass type
    xs ys zs c_csym c_peratom fx fy fz
log logTiO_3d_s_new.data
velocity                all create 100 87384798 rot yes mom yes dist gaussian
#Heating of all region from temperature 100 to 2500K
fix                     ht1 all npt temp 100 2500 0.01 iso 0.0 0.0 0.1 #(5 K/ps)
run                     200000
unfix                   ht1
#Holding of glass region at temperature 1500K with 100 ps holding time
fix                     ho1 all npt temp 2500 2500 0.01 iso 0.0 0.0 0.1
run                     100000
unfix                   ho1
#Cooling of ghaz region from 1500K to 300K temperature with cooling rate 0.5K/ps
fix                     cl1 all npt temp 2500 1300 0.01 iso 0.0 0.0 0.1 #(4 K/ps)
run                     125000
unfix                   cl1
fix                     cl2 all npt temp 1300 700 0.01 iso 0.0 0.0 0.1 #(0.1 K/ps)
run                     1000000
unfix                   cl2
fix                     cl3 all npt temp 700 100 0.01 iso 0.0 0.0 0.1 #(1 K/ps)
run                     300000
```

LAMMPS input script for viscosity calculations

```
units           real
variable        T equal 200.0       # run temperature
variable        Tinit equal 250.0   # equilibration temperature
variable        V equal vol
variable        dt equal 4.0
```

```
variable            p equal 400       # correlation length
variable            s equal 5      # sample interval
variable            d equal $p*$s       # dump interval
# convert from LAMMPS real units to SI
variable            kB equal 1.3806504e-23       # [J/K] Boltzmann
variable            atm2Pa equal 101325.0
variable            A2m equal 1.0e-10
variable            fs2s equal 1.0e-15
variable            convert equal ${atm2Pa}*${atm2Pa}*${fs2s}*${A2m}*${A2m}*
   ${A2m}
# setup problem
dimension           3
boundary            p p p
lattice             fcc 5.376 orient x 1 0 0 orient y 0 1 0 orient z 0 0 1
region              box block 0 4 0 4 0 4
create_box          1 box
create_atoms        1 box
mass                1 39.948
pair_style          lj/cut 13.0
pair_coeff          * * 0.2381 3.405
timestep            ${dt}
thermo              $d
# equilibration and thermalization
velocity            all create ${Tinit} 102486 mom yes rot yes dist gaussian
fix                 NVT all nvt temp ${Tinit} ${Tinit} 10 drag 0.2
run                 8000
# viscosity calculation, switch to NVE if desired
velocity            all create $T 102486 mom yes rot yes dist gaussian
fix                 NVT all nvt temp $T $T 10 drag 0.2
#unfix              NVT
#fix                NVE all nve
reset_timestep 0
variable            pxy equal pxy
variable            pxz equal pxz
variable            pyz equal pyz
fix                 SS all ave/correlate $s $p $d &
                    v_pxy v_pxz v_pyz type auto file S0St.dat ave running
variable            scale equal ${convert}/(${kB}*$T)*$V*$s*${dt}
variable            v11 equal trap(f_SS[3])*${scale}
variable            v22 equal trap(f_SS[4])*${scale}
variable            v33 equal trap(f_SS[5])*${scale}
thermo_style custom step temp press v_pxy v_pxz v_pyz v_v11 v_v22 v_v33
run                 100000
variable            v equal (v_v11+v_v22+v_v33)/3.0
variable            ndens equal count(all)/vol
print               "average viscosity: $v [Pa.s] @ $T K, ${ndens} atoms/A^3"
```

5 Grain Boundary Engineering Using MD Simulation

5.1 INTERFACES IN METALS AND THEIR IMPORTANCE

Grain boundaries are interfaces commonly found in polycrystalline materials that have a significant impact on their properties. GBs play a crucial role in various physical processes, such as diffusion, segregation, cavitation, corrosion, and decohesion. The presence of GBs can significantly affect the mechanical strength of a material [Watanabe and Tsurekawa, 1999]. Moreover, GBs are ideal for studying the geometrical aspects of structure–property relationships. Unlike bulk materials, the interface's three distinct geometrical aspects can be analyzed without the interference of various material combinations or dimensional interface parameters. GB energy is considered a vital factor influencing various GB properties, including impurity segregation, GB mobility and fracture, GB diffusion, and cavitation. Therefore, understanding the correlation between GB structure and energy provides valuable insights into complex structure–property relationships. Impurity segregation and the accumulation of alloying elements at GBs can lead to brittle cracking at low temperatures. Similarly, the segregation of alloying elements can positively or negatively affect fracture during creep caused by cavitation at GBs [Hondros and Seah, 1977].

Grain boundary segregation is a phenomenon that occurs at interfaces between grains in materials, where certain elements preferentially accumulate or segregate. This process has significant industrial and technological implications, as it can influence the properties of materials such as strength, ductility, and corrosion resistance. To better understand grain boundary segregation, researchers have developed various techniques for measuring it, such as the interfacial energy approach, X-ray analysis, transmission electron microscopy, and atom-probe tomography [Kobayashi et al., 2009; Raabe et al., 2014]. These techniques have their own unique contributions and can provide insight into the chemical system dependence and response to environmental parameters of the segregation. Theories of segregation have also been developed, which use analogues of gas adsorption on free surfaces. For example, McLean's segregation theory is analogous to Langmuir adsorption, while Fowler's theory is an extension for self-interacting segregants [Seah, 1980]. These theories can be applied to both limited and unlimited numbers of sites and have been extended to ternary systems.

Recently, there has been growing interest in utilizing grain boundary segregation for material design and engineering. This approach, known as segregation engineering, involves deliberately manipulating the segregation of certain elements to control the properties of the material (as shown in Figure 5.2). Scientists can design

DOI: 10.1201/9781003323495-5

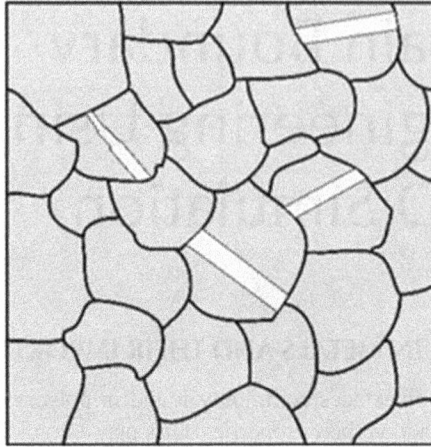

FIGURE 5.1 A typical microstructure of a metallic specimen showing multiple grains (gray) and a network of grain boundaries (dark lines).

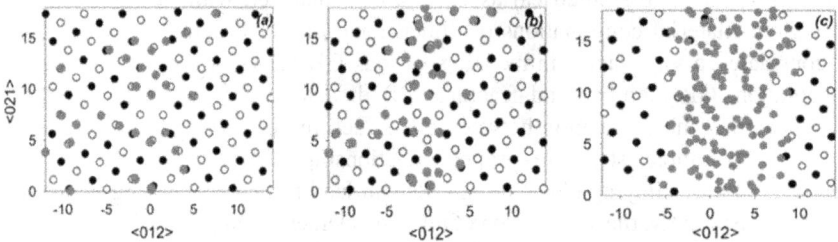

FIGURE 5.2 The schematic shows the solute segregation at the grain boundaries of a metallic system. The red dots denote the solute particle that is first randomly distributed and then subsequently segregated at the GBs [Karkina et al., 2019].

interfaces in metallic alloys by taking into account several factors, including the segregation coefficient of the decorating element, its influence on interface cohesion, energy, structure, and mobility, its diffusion coefficient, and the free energies of competing bulk phases, precipitate phases, or complexions. Through careful consideration of these factors, researchers can identify matrix–solute combinations that are suitable for designing interfaces in metallic alloys.

The practicality of segregation engineering makes it particularly useful for large-scale manufacturing. A modest diffusion heat treatment can usually realize the desired segregation, making it a valuable tool for improving the properties of materials. Overall, grain boundary segregation is a complex phenomenon with significant implications for materials science and engineering. By developing a better understanding of the underlying theories and measurement techniques, and by utilizing segregation engineering, researchers can improve the properties of materials and advance industrial applications. Grain boundary cracking is another crucial issue

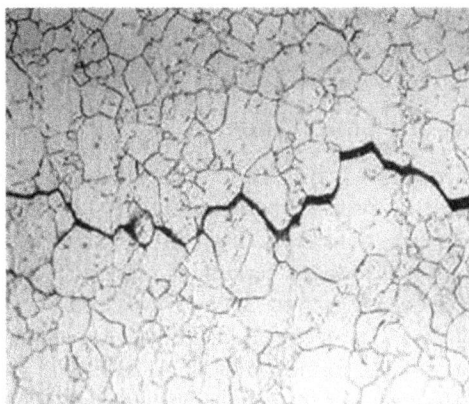

FIGURE 5.3 This micrograph displays intergranular cracking in an Inconel heat exchanger tube, with the crack propagating along the grain boundaries [Metallurgical Technologies, Inc. (MTi)].

encountered in intermetallic compounds, which would otherwise be attractive structural materials for high-temperature applications. Intermetallic compounds have the potential to be excellent structural materials for high-temperature applications. This problem is particularly prevalent in complex compounds, where brittleness can arise due to an inadequate number of slip systems. As a result, extensive plastic deformation can develop, leading to material failure. In L12 compounds like Ni_3Al, which are based on FCC structures, intergranular fracture occurs readily despite having adequate deformation modes. It is believed that brittleness at grain boundaries is an inherent feature of these compounds. However, doping these materials with boron can significantly enhance their ductility by segregating them to the grain boundaries. The atomic structure of grain boundaries is critical to understanding the physical mechanisms underlying various boundary phenomena since they occur in a narrow region of only a few atomic spacings where two grains join. High-resolution electron microscopy and synchrotron X-rays have recently made it possible to observe atomic configurations in grain boundaries directly. This chapter provides an overview of the different phenomena associated with grain boundaries in metallic materials, including diffusion, decohesion, segregation, cavitation, and brittle fracture at low and high temperatures. We examine the atomic structure of grain boundaries and the role of impurities segregated to them in cohesion and diffusion. Moreover, we explore the impact of solute atom doping on the ductility of materials using MD simulations. Finally, we emphasize the importance of understanding grain boundary phenomena for the advancement of structural materials.

5.2 TYPES OF GRAIN BOUNDARIES AND INTERFACES

Before delving into the analysis of grain boundary properties for the metallic systems, it is important to understand how grain boundaries are characterized and identified and the different types of grain boundaries that can be categorized based on thickness, misorientation, and other parameters, and these discussions are summarized in the subsequent subsections.

5.2.1 COINCIDENCE SITE LATTICE THEORY

Coincidence site lattice (CSL) theory is a fundamental concept in the study of grain boundaries, which are the interfaces between two crystalline grains in a polycrystalline material. When two grains meet, they can have an infinite number of orientations relative to each other. However, certain orientations result in the lattice points of one grain coinciding exactly with the lattice points of the other. These are known as coincidence orientations, and the resulting structure is called the coincidence site lattice. In two dimensions, the formation of a CSL is easy to see when rotating two lattices on top of each other. For instance, a hexagonal lattice rotated by a certain angle with respect to another hexagonal lattice results in the development of a moiré pattern. The coincident points in the lattice form a superstructure known as the coincidence site lattice. The significance of CSL theory in the study of grain boundaries is evident in twin boundaries ($\Sigma 3$) and other GBs such as $\Sigma 5$, $\Sigma 7$, and $\Sigma 9$, which are the most-studied type of grain boundary. These CSL GBs are formed when two grains are mirror images of each other and are oriented at a specific angle, resulting in a CSL (Figure 5.4). CSL theory has shown that these boundaries can have lower energy and greater stability than other boundaries, leading to unique properties in materials.

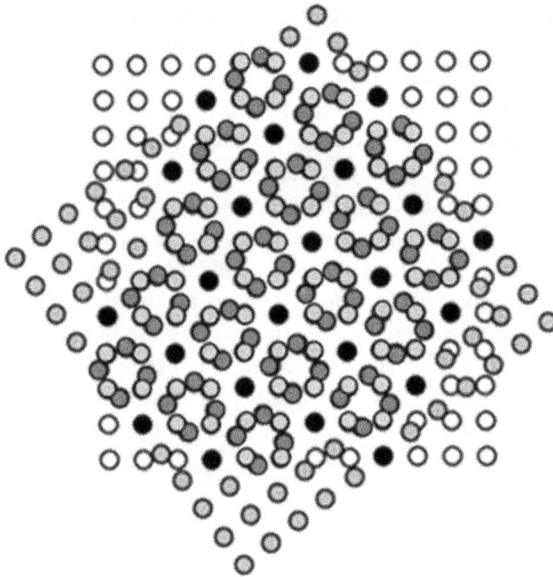

FIGURE 5.4 Schematic representation of overlapping of two crystallographic planes to form $\Sigma 5$ CSL grain boundaries.

The Σ value represents the relationship between two grains in a grain boundary, and this relationship is unambiguous, despite being difficult to identify in some cases. For example, in the case of two orthorhombic or triclinic lattices, it may be challenging to determine the Σ value. In the case of a twin boundary, the Σ value is always 3, whether it is a two-dimensional or three-dimensional lattice. Therefore, twin boundaries are referred to as $\Sigma 3$ boundaries. A $\Sigma 1$ boundary would indicate a near-perfect crystal with only minor misorientations. These misorientations are referred to as small-angle grain boundaries and are typically categorized as $\Sigma 1$ boundaries. Since the Σ value is always an odd number, the twin boundary has the most special coincidence orientation, with the largest number of coinciding lattice points. The next most special coincidence orientation is represented by the $\Sigma 5$ relation, which can be visualized by rotating two square lattices on top of each other in the case of a two-dimensional lattice.

One peculiar fact about CSL boundaries is that they always have odd sigma (Σ) values. To understand the reason behind this phenomenon, it is essential to examine the generation of grain boundaries through rotations around a suitable axis, which results in a rotation angle of γ. In the case of a square lattice, one rotation can produce all possible CSL orientations. Out of all the rotations, some will produce CSL structures, which can be easily conceived. By setting the origin at the apex of a lattice, we can create a CSL orientation by looking at lattice points with coordinates $(x, -y_0)$, which can be expressed as $(n, -1)$ after setting x_0 and y_0 equal to 1 and then rotating

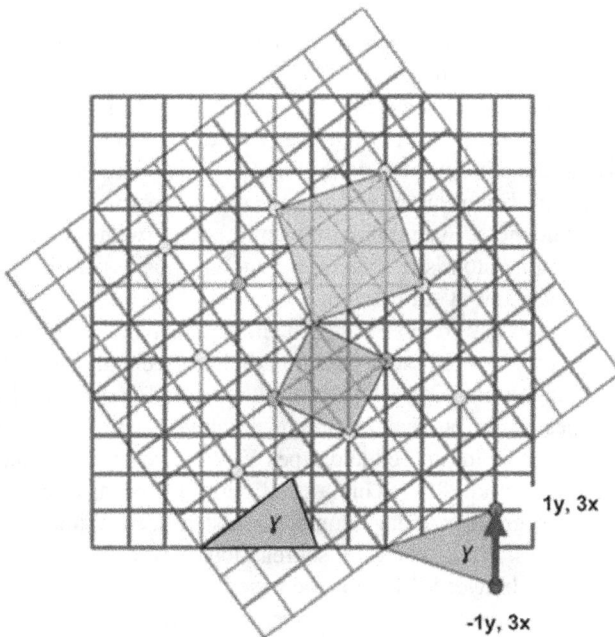

FIGURE 5.5 Schematic representation of two crystal planes to understand the lattice rotation that demonstrates the odd values of Σ in the CSL theory.

the crystal to change the y-coordinate from -1 to $+1$. This CSL lattice's Σ value can be determined by dividing its area by the area of a unit cell of the lattice. This can be generalized for CSL sites generated by shifting the point $(n_x, -y)$ to the $(n_x, +y)$ position.

Odd values of n result in an additional coincidence point in the center of the lattice, which is not present for even values of n. As a result, odd-numbered lattices have a smaller lattice constant than even-numbered lattices. Therefore, the fact that CSL boundaries always have odd Σ values can be attributed to the fact that the required rotation angles result in additional coincidence points in odd-numbered lattices. While a full mathematical proof requires the use of O-lattice theory, this explanation provides an intuitive understanding of this geometric phenomenon.

5.2.2 GRAIN BOUNDARY COMPLEXIONS

In the field of kinetic engineering, a new concept has been introduced called grain boundary or interface complexion. This concept emphasizes the important role that interfaces play in transport kinetics and how they can possess a diverse range of characters. Interface properties such as roughness, solute adsorption, pre-melted films, and changes in crystallographic symmetry can differ significantly. These properties are characterized by a particular combination of thermodynamic variables, interface inclinations, and misorientations, and they exhibit thermodynamic stability. Unlike bulk phases, these interface properties cannot be isolated from their neighboring bulk phases; hence, they do not meet the conventional definition of a phase. A proposal has been made to extend the Gibbs definition of phase to accommodate these features, which are now referred to as interface complexions. These complexion states possess thermodynamic stability and are characterized by specific sets of intensive thermodynamic variables, interface inclinations, and misorientations. Since interface complexions are inseparable from their adjoining bulk phases, they do not conform to the conventional definition of a phase. Interface complexions are associated with equilibrium quantities such as excess volume, entropy, adsorption, and equilibrium thickness. Even if two distinct interfaces with the same complexion may not have the same atomistic structure, they will possess similar characteristic thermodynamic quantities at equilibrium.

The term "complexion" refers to stabilized grain boundary phases that differ from bulk phases yet are highly dependent on their adjacent grains and cannot exist independently [Tang et al., 2006]. Dillon et al. classified grain boundary complexions into six types based on their structure, thickness, and segregation patterns, with types I–IV considered ordered and types V–VI disordered [Dillon et al., 2007; Jud et al., 2005; Nie et al., 2017; Gupta et al., 2007]. Recent research has shown that the structure of grain boundary complexion significantly impacts a material's mechanical properties. As an instance, segregation of Bi atoms into a nickel interface in the form of a bilayer leads to a decrease in ductility, while the presence of intergranular films enhances both strength and plasticity in ceramic materials with nanocrystalline structure [Luo et al., 2011]. Experimental study of ordered complexions is challenging due to their atomic nature, making atomistic simulations a useful tool to gain a detailed understanding of their structural changes at the atomic level.

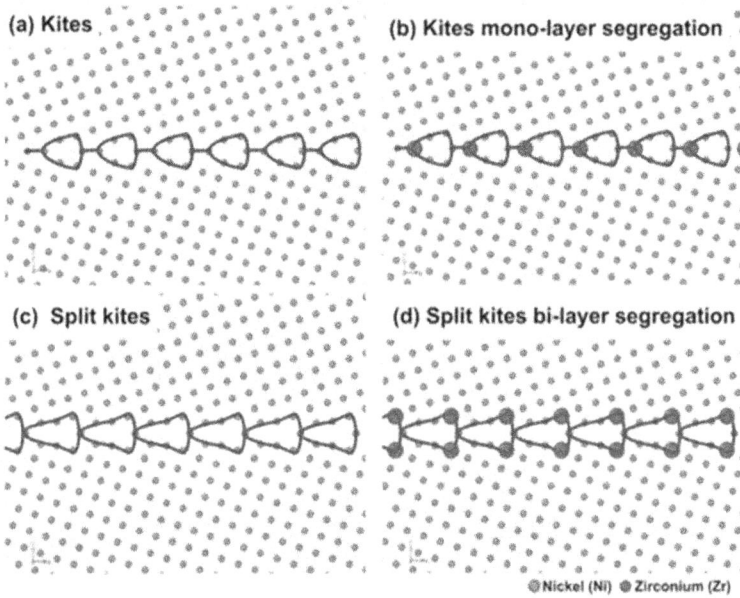

FIGURE 5.6 Atomic snapshots of grain boundary complexions for $\Sigma5$ CSL GB structure. The simulated structures shown here are ordered GB structures with solute segregation (blue Zr atoms).

Most studies on complexions focus on their stability, transitions, and the effect of segregation on their properties due to thermal treatment. However, there are only a few reports studying the effect of complexion on a material's deformation behavior. Understanding the impact of grain boundary complexion on mechanical properties is crucial for improving the performance and functionality of materials in various engineering applications.

In a case study, we have demonstrated the influence of grain boundary complexion on the creep deformation behavior of nickel bicrystal specimens that was investigated using molecular dynamics simulations in a simulated bending creep test [Reddy and Pal, 2018a]. The deformation process revealed strain bursts during low temperature (500 K) bending creep deformation (Figure 5.7). The investigation revealed that specimens with a GB complexion of kite monolayer segregation exhibited stability at low temperatures, while specimens with a split-kite GB complexion were stable at higher temperatures up to 900 K. However, when the creep temperature was further elevated to 1100 K and 1300 K, the split-kite GB complexion became unstable, leading to early failure of the specimen. Moreover, during the bending creep process, the split-kite bilayer segregation and normal kite GB complexions showed localized increases in elastic modulus at 1100 K and 1300 K, respectively, due to the formation of interpenetrating icosahedral clusters. These findings provide valuable insights into the significance of GB complexion in creep deformation behavior. Understanding GB complexions is essential for designing and developing advanced materials with enhanced mechanical properties. It is noteworthy that the role of GB complexion in

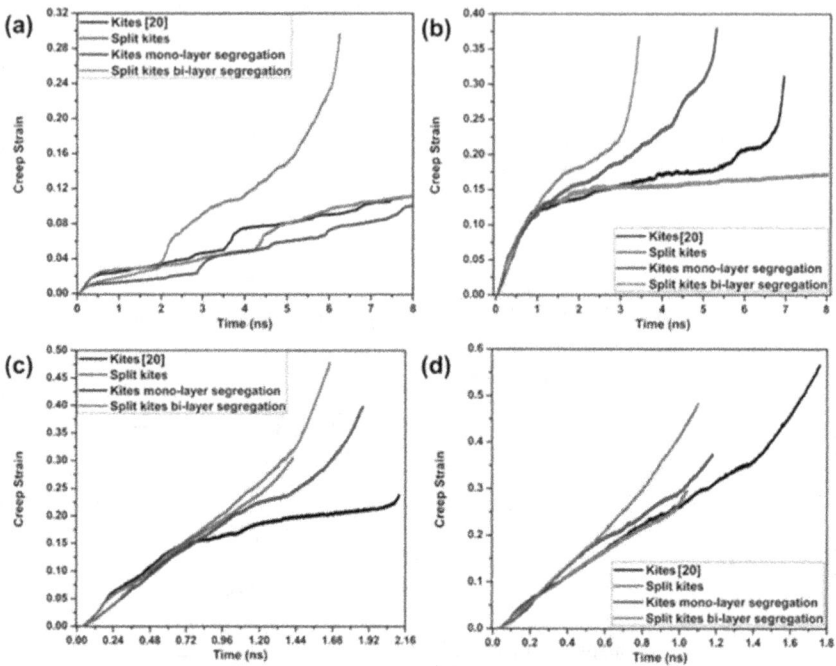

FIGURE 5.7 Creep strain plotted against time for specimens with different grain boundary complexions at various temperatures: (a) 500 K, (b) 900 K, (c) 1100 K, and (d) 1300 K. The grain boundary complexions include normal kite, split-kite, kite with monolayer Zr segregation, and split-kite with bilayer Zr segregation.

the mechanical behavior of materials is not limited to creep deformation alone. For example, in metallic alloys, GB complexions have been found to affect hydrogen embrittlement, stress corrosion cracking, and fatigue behavior. Therefore, exploring and manipulating GB complexions could be a promising strategy for tailoring the properties of advanced materials for a wide range of applications.

In another study, MD methods were used to investigate the creep deformation behavior of nickel bicrystal specimens, focusing on the role of grain boundary complexion [Reddy and Pal, 2018b]. The simulated bending creep test revealed strain burst phenomena during low-temperature (500 K) bending creep, with strain burst occurrence depending on the ease of GB migration. Analysis showed that specimens with kite monolayer segregation GB complexion were stable at low temperatures, while specimens with split-kite GB complexion were stable at higher temperatures (900 K). However, at elevated temperatures (1100 K and 1300 K), the split-kite GB complexion became unstable and led to early specimen failure. During the deformation process, interpenetrating icosahedral (I-ICO) clusters were observed in specimens with split-kite bilayer Zr segregation GB complexion for deformation loads of 8 and 9.6 pN (Figure 5.8). The formation of these clusters reduced the atomic volume and increased structural stability, contributing to the enhancement of creep plasticity

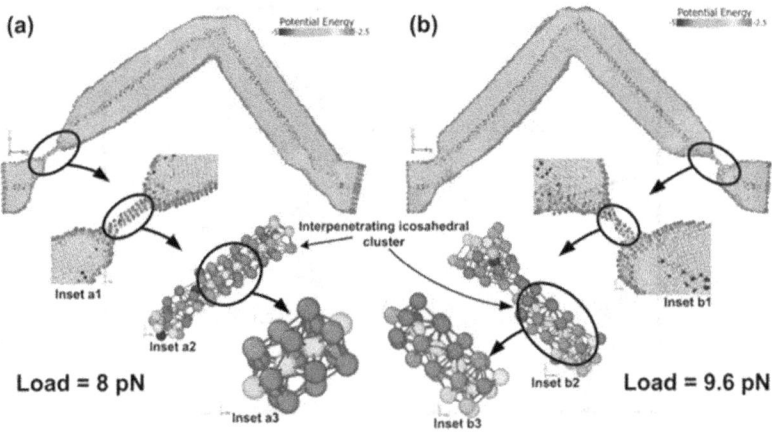

FIGURE 5.8 Atomic snapshot of potential energy per atom of specimen with split-kite bilayer Zr segregation GB complexion before the fracture of the specimen for (a) 8 pN deformation load and (b) 9.6 deformation pN load. Insets show the formation of interpenetrating icosahedral cluster.

in the specimen. This was due to the high local average elastic rigidity resulting from the I-ICO cluster's structural stability and low average potential energy. The segregated bilayer Zr atoms pinned the grain boundary and resisted GB migration, leading to a decrease in the potential energy of the GB complexion. Overall, this study sheds light on the importance of GB complexion in creep deformation behavior and provides insights into the role of I-ICO clusters in enhancing creep plasticity.

5.2.3 Low-Angle and High-Angle GBs

In the field of materials science, there is a common understanding that low-angle grain boundaries (LAGBs) refer to those with a misorientation angle of less than approximately 15 degrees. Conversely, high-angle grain boundaries (HAGBs) are interfaces with a misorientation angle greater than approximately 15 degrees (Figure 5.9). LAGBs are typically created during the growth of a crystal or during recrystallization. They are often considered more "ordered" than HAGBs, as the misorientation between the grains is relatively small. This implies the atomic arrangement across the LAGB is more similar to that of the bulk crystal, which can be beneficial for certain material properties. One interesting feature of LAGBs is that they can act as barriers to dislocation motion, which can affect the mechanical properties of a material. Dislocations are defects in a crystal structure that can cause plastic deformation (permanent changes in shape) when they move through the material. In some cases, the presence of an LAGB can prevent dislocations from moving through the crystal, effectively "locking" the material and making it more difficult to deform. This can be useful in some applications, such as in the manufacture of thin films, where it is desirable to have a material that is resistant to plastic deformation.

In contrast, HAGBs are often considered more "disordered" than LAGBs. This is because the misorientation angle between the grains is larger, which means that

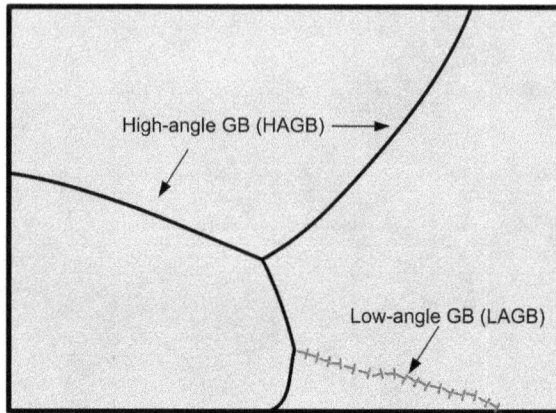

FIGURE 5.9   A schematic model that illustrates the formation of HAGBs and LAGBs.

the atomic arrangement across the HAGB is more different from that of the bulk crystal. HAGBs are typically created during the processing of a material, such as during rolling or forging when the grains are deformed and stretched. They can also form when two crystals with different orientations grow together, such as in a poly-crystalline material. One important feature of HAGBs is that they can act as sources and sinks for dislocations. When a dislocation encounters an HAGB, it can either be absorbed by the HAGB (acting as a sink), or it can generate a new dislocation on the other side of the HAGB (acting as a source). This means that HAGBs can play a significant role in the plastic deformation of a material, as they can facilitate the movement of dislocations through the crystal. Another interesting feature of HAGBs is that they can form "twin boundaries" under certain conditions. Twin boundaries are special types of HAGBs that have mirror symmetry across the boundary plane and are categorized as Σ3 GBs. They can enhance several material properties such as ductility and strength by acting as hardening agents while allowing the material to deform plastically.

5.2.4   AMORPHOUS INTERGRANULAR FILMS

There are several ways to describe grain and hetero-phase interfaces, but among them are amorphous intergranular films, which belong to the class of disordered GB complexions. AIFs are unique because of their constant thickness of 1–2 nm, regardless of the orientation of the bounding grains. However, the thickness is dependent on the composition of the adjacent metallic system, as shown in Figure 5.10. The AIF structure plays a crucial role in determining the properties of ceramic materials, including their resistance to oxidation, creep, and fracture. The AIF is therefore a key factor in controlling the behavior of ceramics such as Si-based ceramics, alumina (Al_2O_3), and others, where the grain boundary region may have an amorphous film of approximately 1–2 nm thickness. These films have a unique property of having a nearly constant thickness that is independent of the orientation of the bounding

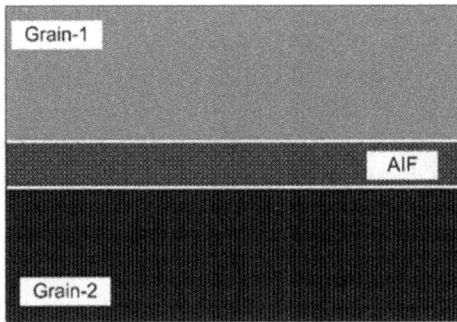

FIGURE 5.10   High-resolution micrograph from a Lu–Mg-doped Si_3N_4 sample showing the presence of an amorphous intergranular film.

grains but is dependent on the composition of the ceramic material. The structure and presence of AIFs play a vital role in determining the overall properties of the ceramic, including fracture, creep, oxidation, and electrical behavior. AIFs can be formed through various synthetic routes such as liquid phase sintering, solid-state activated sintering, and crystallization of glass surrounding the crystal. Moreover, even equilibrium-thickness amorphous films on surfaces have been synthesized that are equivalent to AIFs on surfaces. The advent of advanced microscopy techniques, such as high-resolution microscopy, Fresnel contrast imaging, and electron energy loss spectroscopy, has provided significant insights into the structure of AIFs and their interface with the bonding crystals. It has been discovered that the AIF is graded in terms of composition and structure, with a diffuse interface between the amorphous material and the bonding crystals. The structure of the AIF is also different from the bulk glass in the system. The influence of the bonding crystals on the AIF is believed to be a contributing factor to the gradation [Subramaniam et al., 2006]. Despite recent progress, many questions regarding the formation and behavior of AIFs remain unanswered, including their dependence on crystallography and temperature. This overview aims to provide a basic understanding of AIFs for those without specialized knowledge, highlighting significant advancements in the field and identifying unanswered questions.

5.3   GRAIN BOUNDARY ENGINEERING

The development of high-performance nanocrystalline materials has been a focus of research, with grain boundary engineering (GBE) playing a significant role. GBE involves controlling the distribution of the grain boundary character and connectivity at triple junctions, resulting in a high density of grain boundaries and triple junctions. This approach has been successful in improving the strength, hardness, and resistance to intergranular brittleness and fatigue fracture of nickel and nickel alloys with sub micrometer-grained structures. Another approach is to study the GB excess free volume, GB energy, and segregation energy and determine best suitable interfaces for specific operating conditions. This approach has the potential to provide precise control over the grain boundary microstructure and produce materials with unique mechanical properties. Here we present a case study that shows the analysis of GB properties (different CSL boundaries) using MD simulation [Pal et al., 2021].

The thermodynamic and kinetic properties of grain boundaries, such as their energy, mobility, diffusivity, and segregation energy, are significantly influenced by the excess volume of GBs. However, a simplistic approach of reporting the excess volume as an average value for the entire GB fails to capture the complex nature of this property, particularly for kinetic processes that involve only a few atoms at a time. Therefore, atomistic simulations were utilized to examine the spectral nature of atomic excess volume in representative nanoscale Ni and Al samples. The findings of the research suggest that the atomic excess volume and excess energy in the GBs exhibit a skew-normal distribution, which have a weak inverse correlation with each other. This contradicts the commonly reported positive correlation based on average values. The analysis further demonstrated that the correlation between the atomic excess volume and excess energy heavily depends on the GB type and that a universal trend between them is absent. However, a strong correlation was found between atomic excess volume and segregation energy based on a nanocrystalline Al model with Mg impurity. This study suggests that excess volume can be utilized as a structural variable for future thermodynamic modeling of GBs and can provide useful insights into the elemental structure–property correlations in GBs.

The graphs in Figure 5.11 demonstrate distinct trends in the excess volume and excess energy of various grain boundaries. It is observed that all the $\Sigma 3$ GBs exhibit a negative slope, indicating that an increase in excess volume results in a decrease in excess energy. In contrast, the (111) twist GB displays a positive slope, indicating that excess energy increases with excess volume. The graphs for a thermal annealed (TA) GB and an athermal (A) GB both exhibit bimodal distributions, with strong correlations between excess volume and excess energy for all atoms. The TA GB also displays a trimodal distribution, with peaks at -0.18, 0.51, and 1.21 Å^3. The excess volume spectrum for non-thermal GBs at 100 K can be interpreted as a bimodal distribution, with the first peak near zero and the second peak at a larger positive excess volume value. Multiple peaks are observed in the excess energy distribution for thermal and athermal GBs, with most atoms having positive excess energy values and those with excess volume close to zero having positive excess energy values. The excess volume distribution can roughly predict GB mobility, but a direct correlation between the excess volume spectrum and mobility is difficult to establish. These findings suggest that compared to the TA $\Sigma 3$ GB, A or AT $\Sigma 3$ GBs have either significantly smaller positive excess volume or a significantly smaller population of atoms with positive excess volume. This supports previous studies that suggest the movement of AT GBs is slowed down due to the increased duration of stagnant atoms.

Aside from investigating the correlation between excess volume and excess energy on a per-GB basis, we also evaluated the trend for different types of GBs. Our results showed that low-Σ GBs (which possess symmetrical and repeatable unit structures) displayed a strong correlation between excess energy and excess volume, exhibiting a steep negative slope and high R-squared values. However, as the Σ value increased and GBs became more distorted, the correlation weakened, resulting in lower R-squared values and higher slope values. We also observed a unique trend for (111) twist GBs,

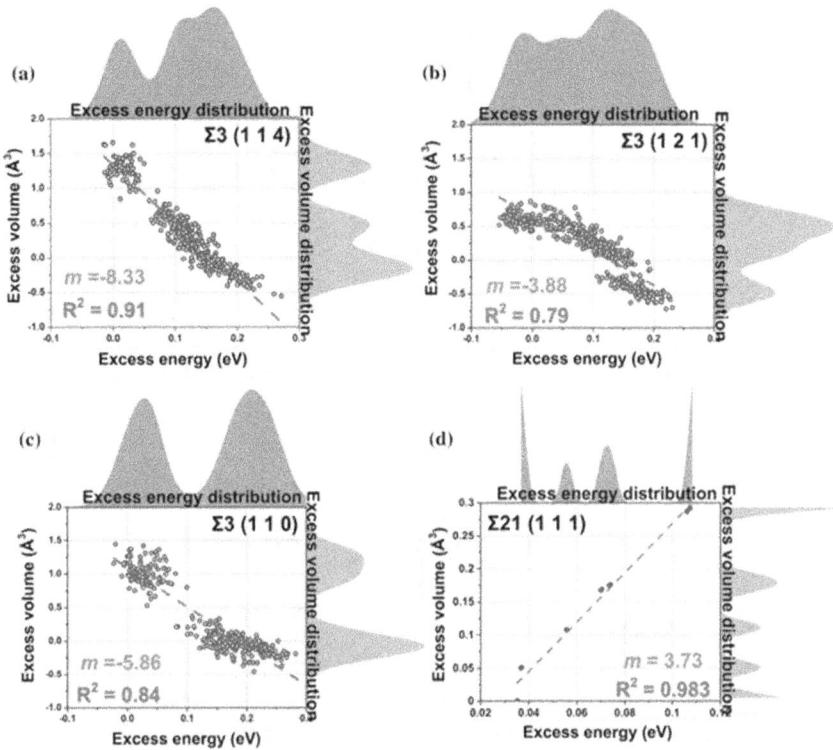

FIGURE 5.11 The correlation between atomic excess free volume and excess energy in various types of Σ3 grain boundaries at 100 K. The four plots illustrate the distribution of excess free volume (blue) and excess energy (pink) for (a) a thermal GB (PID 18-Σ3(114)), (b) an athermal GB (PID 4-Σ3(121)), (c) an anti-thermal GB (PID 5-Σ3(110)), and (d) a (111) twist GB (PID 31-Σ21(111)).

which exhibited a positive slope value, unlike all other GB types. While we currently lack a clear explanation for this phenomenon, it may be connected to the results of our atomic von Mises stress analysis for (111) twist GBs. Our findings suggest that the local atomic environment is a crucial factor in determining excess energy and excess volume and that more sophisticated descriptors like machine learning–based SOAP may be required to accurately forecast these properties.

It is essential to highlight that one of the Σ3 GBs scrutinized in this investigation displays a considerably lower R-squared value in comparison to the others (Figure 5.12(a)). Further scrutiny revealed that this specific GB has a faceted boundary characterized by a "hill-and-valley" morphology, and a significant portion of the boundary constitutes a facet of the coherent twin boundary. The excess energy and excess volume distributions of the GB atoms are assessed and illustrated in Figure 5.13. The excess volume distribution plot reveals a single peak proximate to zero and a minor peak close to 0.2 Å^3. Conversely, the excess energy plot shows numerous peaks, with the most prominent peak close to zero and minor peaks at

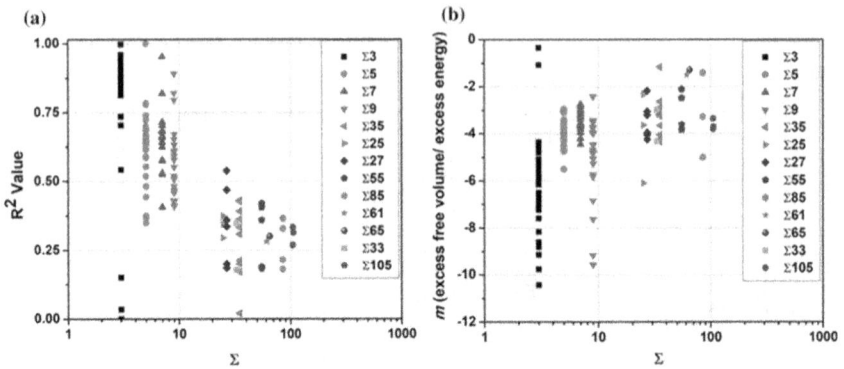

FIGURE 5.12 Distribution of R-squared values and slope (m) values for low- and high-Σ grain boundaries.

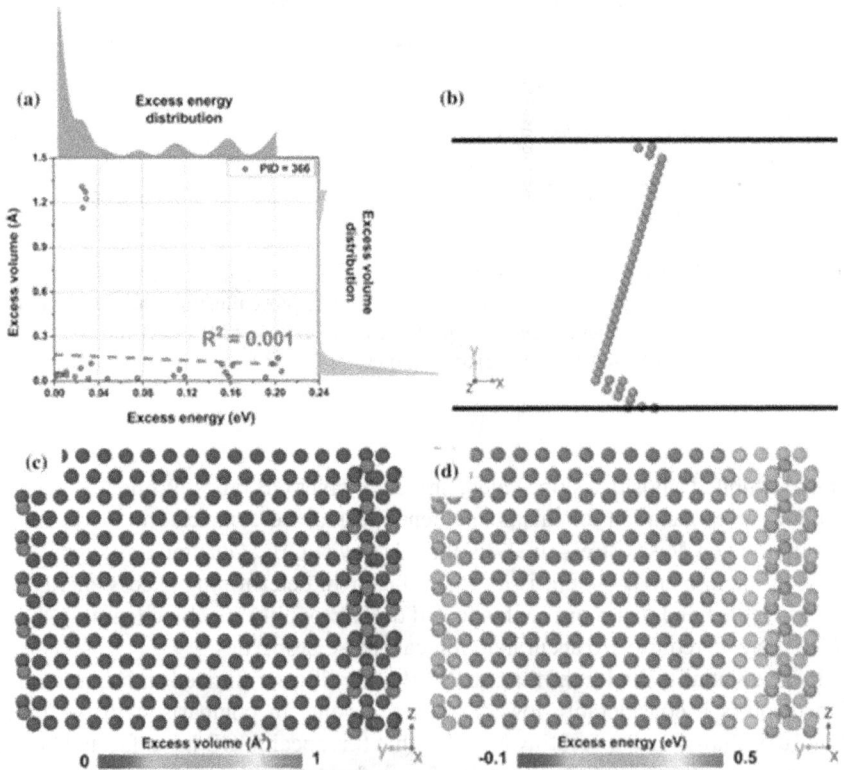

FIGURE 5.13 (a) Excess energy versus excess volume correlation plot and corresponding distributions of excess free volume and excess energy for specimen with low R-squared value. (b) Atomic snapshot highlighting the GB atoms, where atoms on the coherent twin facet are colored in red. (c) Atomistic configurations of the GB, color-coded based on their excess volume. (d) Atomic excess energy of the GB.

higher energy values. This can be attributed to the structure of the faceted boundary, which shifts in the GB plane, leading to a change in the GB energy (Figure 5.13(b)). This phenomenon has been extensively discussed and supported in the literature. Therefore, when plotting the excess energy versus excess volume scatter plot, it is evident that the excess volume values are confined, whereas the excess energy values scatter across the spectrum, resulting in a significantly low R-squared value (Figure 5.13(c)).

5.4　DESIGNING AND ANALYZING METALLIC GBs USING LAMMPS

In this section, we demonstrate the use of LAMMPS in designing grain boundary structures (bicrystal specimens) and simulating mechanical deformation tests to study the GB-related behavior in the specimens [Pal et al., 2020]. In general, it is known that the full potential of nanocrystalline metals as advanced structural materials is limited due to GB movements and growth. To address this issue, our research aimed to examine the influence of Zr segregation on the mechanical properties of a Ni grain boundary under shear deformation and high-temperature bending creep. Specifically, we investigated the impact of segregated Zr on the migration of GBs and its resulting strengthening effect in nanocrystalline metals. Through molecular dynamics simulations, we analyzed the deformation of a Ni bicrystal specimen with a symmetric Σ5 GB, both with and without segregated Zr. Our findings suggest that GB segregation of up to 0.4 at. % can effectively pin the CSL boundary, leading to an increase in shear strength. This trend is also observed in the case of bending creep deformation, where creep resistance is enhanced up to 0.4 at. %. However, excessive GB segregation can cause destabilization of the local atomic structure, leading to a reduction in GB stiffness.

The impact of Zr segregation on the Ni bicrystal specimen during low-temperature (100 K) and high-temperature (900 K) shear deformation was investigated using molecular dynamics simulation. The simulation results indicate that segregation of

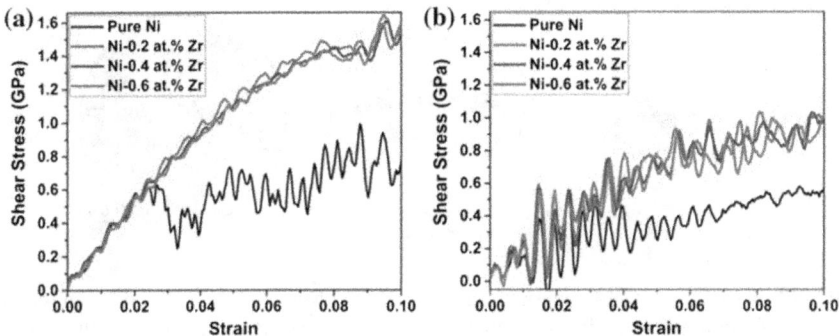

FIGURE 5.14 Stress–strain curves of a pure Ni bicrystal and a Ni-Zr bicrystal during shear deformation at 100 K (a) and 900 K (b).

Zr atoms to the grain boundary significantly increases the yield stress of the specimen during low-temperature shear deformation. The stress–strain plot shows that the onset of plastic deformation is delayed due to effective pinning of the GB, suppressing GB migration. On the other hand, the clean CSL boundary specimen exhibits lower strength during high-temperature shear deformation compared to the GB segregated specimens. At higher temperatures, the internal energy increases, facilitating smooth propagation of dislocations, resulting in lower strength and higher ductility. Large serrations in the stress–strain plot indicate recurrent dislocation generation-annihilation processes.

Furthermore, Figure 5.15 presents the atomistic mechanism of shear-coupled GB migration through disconnections. The lower grain moves in the negative x-direction, causing distortions in the GB and the formation of metastable structures that accommodate the initial plastic strain. The metastable structure transforms into a perfect kite structure, creating a step in the GB region called a disconnection. This continuous shear-driven process causes the GB to migrate toward the upper surface of the specimen. The understanding of this process is essential to gain insight into the mechanical behavior of nanocrystalline metals and to develop advanced structural materials with improved properties.

In Figure 5.16, we present a quantitative analysis of the displacement of the grain boundary (ΔGB) over time to investigate the migration of clean and Zr-segregated boundaries. The displacement of atoms at the grain boundary region is calculated for various time steps to determine ΔGB. Our analysis focuses on the specimens with clean and 0.4 at. % Zr-segregated boundaries, which display the lowest and highest creep resistance, respectively. The results indicate that the clean boundaries exhibit

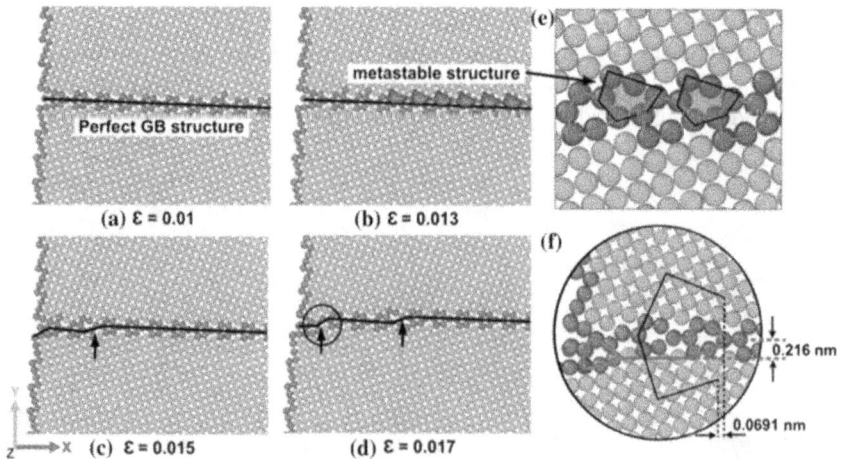

FIGURE 5.15 (a–d) CNA snapshots showing the mechanism of grain boundary migration through disconnections during low-temperature (100 K) shear deformation of the pure Ni bicrystal specimen. The metastable grain boundary structures formed during the process are shown in (e), and a magnified snapshot of the disconnection step observed in (d) is presented in (f).

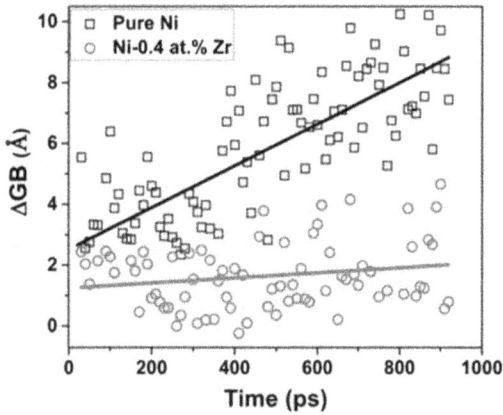

FIGURE 5.16 Grain boundary displacement (ΔGB) versus time during the bending creep process for Ni bicrystal specimens with clean and 0.4 at. % Zr segregated to the GB.

a significantly steeper slope compared to the Zr-segregated boundary, suggesting that the presence of Zr atoms impedes GB migration by pinning the $\Sigma5$ CSL boundary's kite structure. This leads to reduced migration during high-temperature creep deformation. Previous studies using molecular dynamics have reported that Zr segregation can decrease the grain boundary energy of bicrystal specimens [Reddy and Pal, 2018c; Dillon et al., 2007], which enhances their resistance to boundary motion during bending creep deformation.

5.5 EXAMPLE LAMMPS INPUT CODES

LAMMPS script to create bicrystal with $\Sigma5$ CSL GBs

```
# LAMMPS Input File for Grain Boundaries
# Mark Tschopp, Dec2009
# This file will generate a single Sigma5(310) STGB
#---------------Initialize Simulation---------------
clear
units           metal
dimension       3
boundary        p p p
atom_style      atomic
#--------------Create Atomistic Structure------------
lattice     fcc 3.52
region          whole block 0.000000 600.000000 0.000000 100.000000
    0.000000 100.000000 units box
create_box      1 whole
region          upper block INF INF 0.000000 50.000000 INF INF units box
lattice     fcc 3.52 orient x 1 3 0 orient y -3 1 0 orient z 0 0 1
create_atoms    1 region upper
region          lower block INF INF 50.000000 100.000000 INF INF units box
lattice     fcc 3.52 orient x 1 -3 0 orient y 3 1 0 orient z 0 0 1
```

```
create_atoms      1 region lower
group             upper type 1
group             lower type 1
#--------------Define Interatomic Potential--------------
pair_style        eam/fs
pair_coeff        * * Ni1_Mendelev_2012.eam.fs Ni
neighbor          2.0 bin
neigh_modify      delay 10 check yes
#--------------Displace atoms and delete overlapping atoms--------------
displace_atoms    upper move 0 0 0 units lattice
delete_atoms      overlap 1.5 lower upper
#--------------Define Settings--------------
compute           csym all centro/atom fcc
compute           eng all pe/atom
compute           eatoms all reduce sum c_eng
#--------------Run Minimization--------------
reset_timestep    0
thermo            10
thermo_style      custom step pe lx ly lz press pxx pyy pzz c_eatoms
dump              12 all custom 25 NVT_defo_Ni_biaxial*.lammpstrj id type x y z
dump              1 all cfg 25 dump.sig5_minimization_*.cfg mass type xs ys zs
   c_csym c_eng fx fy fz
dump_modify       1 element Ni
min_style         cg
minimize          1e-15 1e-15 100 100
undump            1
#--------------Run Minimization 2--------------
# Now allow the box to expand/contract perpendicular to the grain boundary
reset_timestep    0
thermo            10
thermo_style      custom step pe lx ly lz press pxx pyy pzz c_eatoms
fix               1 all box/relax y 0 vmax 0.001
min_style         cg
minimize          1e-15 1e-15 1000 1000
#--------------Calculate GB Energy--------------
variable minimumenergy equal -4.38550876957302
variable esum equal "v_minimumenergy * count(all)"
variable xseng equal "c_eatoms - (v_minimumenergy * count(all))"
variable gbarea equal "lx * lz * 2"
variable gbe equal "(c_eatoms - (v_minimumenergy * count(all)))/v_gbarea"
variable gbemJm2 equal ${gbe}*16021.7733
variable gbernd equal round(${gbemJm2})
print "GB energy is ${gbemJm2} mJ/m^2"
#--------------Dump data into Data file--------------
reset_timestep 0
dump              13 all custom 25 NVT_defo_Ni_biaxial_data_file*.lammpstrj id
   type x y z
dump              1 all cfg 25 dump.al_sig5_310_*.cfg mass type xs ys zs c_csym
   c_eng fx fy fz
dump_modify       1 element Ni
minimize 1e-15 1e-15 100 100
```

```
undump 1
write_restart restart.al_sig5_310_stgb
print "All done"
```

LAMMPS script for shear deformation

```
#--------------INITIALIZATION--------------
units              metal
echo               both
dimension          3
boundary           s s p
atom_style         atomic
#----------------ATOM DEFINITION--------------
read_data          read_data_Ni_dislocation_no_Zr
#--------------FORCE FIELDS--------------
timestep           0.002
pair_style         eam/fs
pair_coeff             * * Ni-Zr_Mendelev_2014.eam.fs Ni
minimize               1.0e-12 1.0e-12 500 500
thermo             100
region                 19 block INF INF INF 539 INF INF units box
group              lower region 19
region                 29 block INF INF 805 INF INF INF units box
group              upper region 29
group              boundary union lower upper
group              mobile subtract all boundary
log                NiZr_0.0001_100K.data
#fix               1 all nvt temp 900 900 0.01
#run               1000
#unfix             1
compute                1 all stress/atom NULL
compute                2 all reduce sum c_1[1] c_1[2] c_1[3]
variable           stress equal ((c_2[3])/(3*vol))
variable           tmp equal xy
variable           lo equal ${tmp}
variable           strain equal (xy-v_lo)/ly
######################### cycle-1 #########################
compute                new mobile temp
velocity           mobile create 900 887723 temp new
velocity           lower set -0.205 0.0 0.0
velocity           mobile ramp vx -0.205 0.0 y 539 805 sum yes
# fixes
fix                1 all nve
fix                2 boundary setforce 0.0 0.0 0.0
# run
timestep           0.002
thermo             200
thermo_modify          temp new
variable           time equal step*0.002
variable           p1 equal "-pxx/10000"
variable      p2 equal "-pyy/10000"
```

```
variable        p3 equal "-pzz/10000"
variable            p12 equal "-pxy/10000"
variable            p23 equal "-pyz/10000"
variable            p13 equal "-pxz/10000"
variable        fm equal "(v_p2+v_p3+v_p1)/3" ##### Hydrostatic stress
variable        fv equal "sqrt(((v_p2-v_p3)^2+(v_p3-v_p1)^2+(v_p1-v_p2)^2+6*(v_
    p12^2+v_p23^2+v_p13^2))/2)" ######Von Mises Stress
variable        t equal "v_fm/v_fv"
variable        fd equal (((v_p2-v_fm)*(v_p3-v_fm)*(v_p1-v_fm))-(v_p12)^2*(v_
    p3-v_fm)-(v_p13)^2*(v_p2-v_fm)-(v_p23)^2*(v_p1-v_fm)+2*v_p12*v_
    p23*v_p13)####Deviatoric Von Mises stress
# principal stresses
variable        l1 equal "(v_p1+v_p2+v_p3)"
variable            l2 equal "((v_p1)*(v_p2))+((v_p2)*(v_p3))+((v_p1)*(v_p3))-(v_
    p12)^2-(v_p23)^2-(v_p13)^2"
variable            l3 equal "((v_p1)*(v_p2)*(v_p3))-((v_p1)*((v_p23)^2))-((v_p2)*((v_
    p13)^2))-((v_p3)*((v_p12)^2))+2*(v_p12)*(v_p23)*(v_p13)"
variable            A equal "(acos((((2*(v_l1)^3)-9*(v_l1)*(v_l2)+27*(v_l3))/
    (2*((v_l1)^2-(3*(v_l2)))^(3/2)))))/3"
variable        s1 equal "((v_l1)/3)+(2/3)*(sqrt((v_l1)^2-(3*(v_l2))))*cos(v_A)"
variable        s2 equal "((v_l1)/3)+(2/3)*(sqrt((v_l1)^2-(3*(v_l2))))*cos((v_A)+(2*PI)/3)"
variable        s3 equal "((v_l1)/3)+(2/3)*(sqrt((v_l1)^2-(3*(v_l2))))*cos((v_A)+(4*PI)/3)"
#fix            def all print 200 "${strain} ${p1} ${p2} ${p3}" file NiZr_alloy_
    stress_strain_0.0001_100K_tensile.txt
thermo_style    custom step temp vol etotal pyy lx ly lz xy v_strain
dump            1 all custom 200 NiZralloy_defo_dump_0.0001_900K_*.lam-
    mpstrj id type x y z
dump            33 all custom 400 stress_peratom* id type x y z c_1[1] c_1[2] c_1[3]
    c_1[4] c_1[5] c_1[6]
fix             def1 all print 200 "${strain} ${p12}" file NiZr_alloy_stress_
    strain_0.0001_900K_tensile1.txt
run             50000
unfix           def1
unfix           1
unfix           33
unfix           2
```

6 MD Simulation of Composite Material

This chapter deals with the mechanical and deformation behavior of composite materials at nanoscale using molecular dynamics simulation. The roles of metal matrix composites and design of composite materials using LAMMPS are also elucidated in detail in this chapter. The evolution of the deformation behavior of composite materials is explained comprehensively with the aid of atomic snapshots. The LAMMPS input code for a metal matrix composite is included in this chapter.

6.1 IMPORTANCE OF NANOSCALE COMPOSITE STRUCTURE

A composite is formed when two chemically and physically distinct identifiable materials are combined in order to accomplish a material that exhibits superior properties compared to the materials in their pure forms. A composite is heterogeneous at a microscopic scale, but at a macroscopic scale, it is homogeneous. It is a unique class of materials that enables us to enhance specific properties of monolithic materials. In a composite, the enhancement of a specific property can be controlled by the proper selection of an amount and size of a particular reinforcement [Chawla, 2012; Hull, 1996; Bauri and Yadav, 2018; Jacoby, 2004]. Traditional composites comprise micron-sized fillers and allow us to tailor their material properties like electrical and thermal conductivities, density, hardness, wear resistance, ductility, and strength.

Metal matrix composites (MMCs) use metals, alloys, or intermetallic compounds as a matrix that is dispersed with a metallic, ceramic, or organic filler material in the form of particulates, platelets, whiskers, continuous and discontinuous fibers, or any other regular or irregular shape. In MMCs, metals like Al, Mg, Ti, Cu, Fe, alloys, or superalloys are among the various metals used as matrices, and the most frequently used reinforcements are carbonaceous fillers, SiC, TiO_2, Al_2O_3, B_4C, Y_2O_3, Si_3N_4, AlN, TiC, and several others. MMCs have a wide area of applications such as aerospace, defense, automobiles, sports equipment, and various other fields [Srivastava, 2017; Rohatgi et al., 2011]. The processing of MMC is categorized into two groups, ex-situ synthesis and in-situ synthesis. When the reinforcement preparation is done externally and is added to the metal matrix from outside, it is referred to as ex-situ synthesis. In this process, reinforcing particles are made separately prior to the fabrication of the composite and are externally mixed with the matrix. The ex-situ synthesis process has two sub-classifications, solid-state synthesis and liquid-state synthesis. The powder metallurgy (PM) route, chemical vapor deposition technique, and electroplating are some of the types of solid-state processing techniques, whereas melt infiltration and various casting processes like squeeze casting, stir casting, and compo-casting are some of the types of liquid-state processing techniques.

DOI: 10.1201/9781003323495-6

On the other hand, when the reinforcement is formed within the matrix through any chemical reaction like mechanical alloying (MA), reactive hot pressing (RHP), direct-melt reaction, or self-propagating high-temperature synthesis (HTS) and is added to the matrix without having any interaction with the external environment, it is known as in-situ synthesis. The MMCs developed by the in-situ process exhibit more advantages than the MMCs developed by the ex-situ process. In the case of in-situ synthesis, the reinforcement remains thermodynamically stable within the matrix, resulting in an unoxidized, strong, and clean particle-to-matrix interface bonding. Apart from this, a final reinforcing phase obtained in the case of in-situ synthesis is uniformly dispersed within the matrix and results in superior mechanical properties. The in-situ process is also cost effective [Sheibani and Najafabadi, 2007; Gotman et al., 1994; Aikin, 1997]. Figure 6.1 shows the development of a composite using a matrix and reinforcement.

Nanocomposites, a new class of materials, incorporate a small number of nano-sized particles with a high specific surface area as reinforcement. Metal matrix nanocomposites (MMnCs), a promising class of nanocomposites, exhibit excellent physical and mechanical properties, making them suitable for various applications. These reinforcements interact more strongly with the metal matrix, resulting in superior properties of the nanocomposites, including physical, structural, thermal, and mechanical properties. MMnCs reinforced with carbonaceous nanofillers such as multiwalled carbon nanotubes (MWCNTs) and graphene have shown exceptional mechanical and wear performance, even in adverse environmental conditions, surpassing conventional materials. Hybrid nanocomposites, a recent advancement in nanocomposites, have significantly better properties than traditional monolithic materials and nanocomposites. Hybrid composites can incorporate different materials or a combination of two or more forms of reinforcements, such as fibers, particulates, whiskers, and nanotubes. This study focuses on fabricating Al and Cu-based MMnCs reinforced with two carbon allotropes, MWCNTs and exfoliated graphite nanoplatelets (xGnPs), a derivative of graphene.

6.2 MD SIMULATION OF DEFORMATION BEHAVIOR IN METAL MATRIX COMPOSITES

Traditionally, micron-sized reinforcements have been used by researchers to tailor the material properties of composites like strength, hardness, thermal and electrical conductivities, and wear performance. However, with the emergence of nanomaterials, nanocomposites began to be developed with properties far superior compared to the composites that were developed using micron-sized reinforcements. The development of nanocomposites became possible with the availability of techniques to synthesize nanosized reinforcements. The surge in the field of nanotechnology has

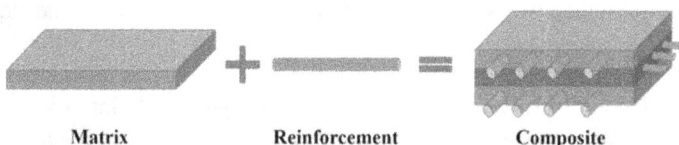

Matrix   Reinforcement   Composite

FIGURE 6.1 Composite made of matrix and reinforcement.

been greatly facilitated by the advent of several analytical techniques that enable us to study the structure of these nanomaterials at atomic resolution. Nanomaterials are capable of providing better reinforcement due to their higher aspect ratio [Omanović-Mikličanin et al., 2020; Casati et al., 2014]. When added as a nanofiller in a matrix, they can significantly improve the property of the nanocomposite.

Nanocomposite materials consist of at least one phase with one, two, or three dimensions in the nanometer range. As the material dimension is at the nanometer level, it creates phase interfaces that are very important for the enhancement of the material property. Nanocomposites are a substitutional way to overcome the limitations of microcomposites and monoliths, and they are becoming the material of the future. Different species of nanosize oxides like Al_2O_3 and Y_2O_3, nitrides like Si_3N_4 and AlN, carbides like SiC and TiC, and borides like TiB_2 have been used as reinforcements in metal matrices. By adding a suitable carbon-based nanofiller like graphite, graphene, graphene oxide (GO), carbon nanotubes, or fullerenes, the wear, electrical, and thermal properties along with the desired mechanical properties of the nanocomposite can be significantly enhanced. Carbonaceous nanomaterials like CNTs and graphene show extraordinary elastic modulus and strength and are ideal nanofillers for developing nanocomposites. MMCs that are reinforced with nanosized materials are called metal matrix nanocomposites. Nanocomposites have an extensive range of applications, as they can overcome the limitations provided by micro-reinforced composites [Haghshenas, 2016; Koilraj et al., 2015; Malaki et al., 2019; Chen, 1995]. Figure 6.2 illustrates the various applications of nanocomposites.

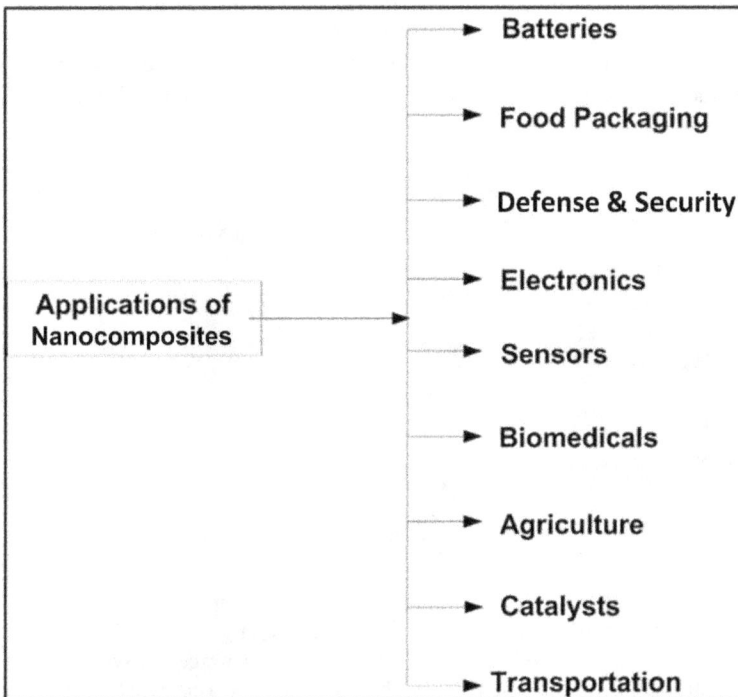

FIGURE 6.2 Nanocomposite applications in various fields.

6.3 DESIGN OF COMPOSITE MATERIALS USING LAMMPS

To create $20 \times 20 \times 20$ nm^3 NC Al specimens, the Voronoi method was used, which involved generating 12 grains with a grain size of 9 nm each using the Atomsk software [Hirel, 2015]. In this study, we inserted armchair-CNTs with (5,5), (15,15), and (30,30) into the NC-Al specimens (Figure 6.3(a)). Choi et al. conducted MD simulation studies on single-crystal Al/CNT composites at 300 K, which showed that the UTS was reached at a strain of 0.18 and fracture strain of 0.45. However, in our study, the UTS of NC Al with (5,5) CNT was reached at a strain of 0.14, and the fracture strain was at least 0.84. The deviation indicates that the position of the CNT with respect to grains and grain boundaries affects the properties. Unfortunately, due to the specimen size, it was not possible to complete such a study. The CNTs in our work passed through grains and grain boundaries, with the x, y, and z orientations being [100], [010], and [001], respectively. The number of atoms in the NC Al specimens without CNT, with (5,5) CNT, with (15,15) CNT, and with (30,30) CNT were 483,274, 486,641, 488,252, and 490,096, respectively. The smallest diameter of the CNTs used in this work was 0.678 nm, while the radius of the aluminum atom was 0.184 nm, suggesting that it was possible to introduce aluminum atoms inside the CNT. To avoid any bias, the inner part of the CNT was filled with aluminum atoms, as the absence of matrix atoms in the CNT would lead to deformation processes that are biased to the matrix-CNT interface and do not represent typical behavior. Our simulation showed that filling the CNT with aluminum atoms prevented void generation at the matrix-CNT interface, which would weaken the composite. The simulation applied periodic boundary conditions in the x-, y-, and z-directions, with a time step of 0.001 ps. An embedded atom method potential and adaptive intermolecular reactive empirical bond order (AIREBO) potential were used for Al-Al and C-C interactions, respectively. The EAM was used to calculate the total energy E_i with the following equation:

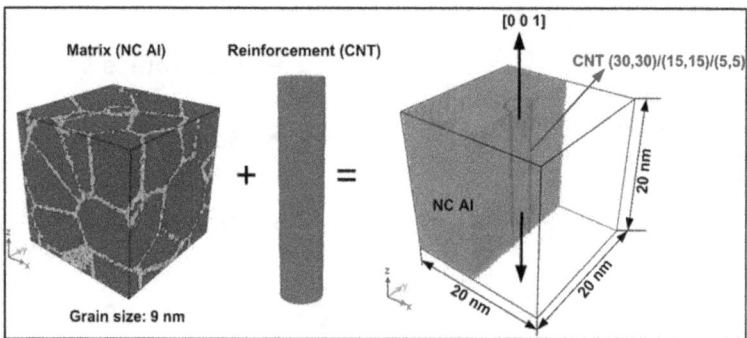

FIGURE 6.3 The starting material for this study is a composite consisting of a carbon nanotube and nanocrystalline aluminum with three different grain sizes. The composite was created by combining the NC Al, which was analyzed using centro-symmetry parameter analysis, with a CNT that was inserted into a hole in the matrix. This initial specimen was used for subsequent simulations and analysis.

$$E_i = F_\alpha \left(\sum_{j \neq i} \rho_\beta \left(r_{ij} \right) \right) + \frac{1}{2} \sum_{j \neq i} \emptyset_{\alpha\beta} \left(r_{ij} \right)$$

where F is the embedded energy (depends on atomic electron density ρ), Φ is the pair potential function, and α and β are the element types of atoms i and j. The AIREBO style computes the energy as

$$E = \frac{1}{2} \sum_i \sum_{j \neq i} \left[E_{ij}^{REBO} + E_{ij}^{LJ} + \sum_{k \neq i, jl \neq i, j, k} \sum E_{kijl}^{TORSION} \right]$$

To model the Al-C interaction, a Lennard–Jones (LJ) 12–6 potential is used based on van der Waals forces, as described by Dandekar et al. [2011]. The cutoff distance for this potential is set to 10.2 Å, which is three times larger than the cutoff distance used for the C-Al interaction to match the cutoff distance of the AIREBO potential. The LJ 12–6 potential is calculated using the following formula:

$$E = 4\epsilon \left[\left(\frac{\sigma}{r} \right)^{12} - \left(\frac{\sigma}{r} \right)^{6} \right]$$

where E is intermolecular potential among the Al and C atoms, σ is the distance at which the inter-atomic potential is zero, r is the distance between two atoms, and ϵ is the potential well depth. The conjugate gradient method was employed for energy minimization, followed by equilibration at 10 K, 300 K, and 681 K (approximately 0.7 times the melting point) using the NPT ensemble, which lasted for 20,000 time steps. The specimens were then subjected to uniaxial tensile loading in the z-direction at a strain rate of 10^{10} s^{-1} in the NVT ensemble to study the atomistic mechanisms of CNT-reinforced NC Al metal matrix nanocomposites during the deformation process [Xiang et al., 2017]. One simulation was conducted at a lower strain rate of 10^9 s^{-1} at 300K to assess the effect of the strain rate. LAMMPS was utilized for MD simulations, while OVITO software was used for post-processing, including dislocation analysis (DXA) and the common neighbor analysis technique to detect dislocation density and stacking faults, respectively. The atomic strain tool of OVITO [Stukowski et al., 2012] was employed to determine the shear strain distribution in the matrix and CNT during deformation. Furthermore, OVITO's Wigner–Seitz defect analysis tool was utilized to analyze vacancies, and the spatial distribution of vacancies was obtained by identifying sites whose occupancy was zero.

6.4 EVALUATION OF DEFORMATION BEHAVIOR AND MECHANICAL PROPERTIES

6.4.1 MECHANICAL BEHAVIOR OF CNT-NC AL NCs

The stress–strain graphs of the NC Al and CNT-NC Al specimens for various temperatures are presented in Figure 6.4, and the corresponding mechanical properties are summarized in Table 6.1. The Young's modulus was determined using

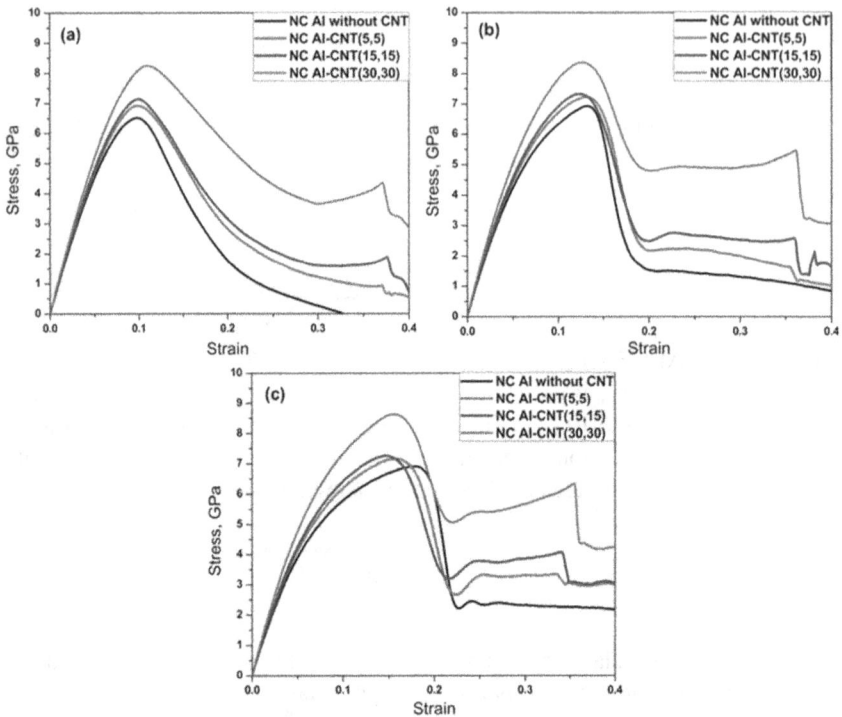

FIGURE 6.4 Stress–strain plots of NC Al without CNT and CNT-reinforced NC Al specimens at different temperatures: (a) 10 K, (b) 300 K, and (c) 681 K.

linear regression analysis, and the yield strength was calculated using the 0.2% off-set method, which is a standard method used in the literature [Jafari et al., 2012; Shimizu et al., 2007]. The results reveal that the Young's modulus and yield strength increase with the increase in CNT diameter at all temperatures. The percentage of modulus improvement (for the NC Al-(30,30) CNT compared to the NC Al without CNT) is about 14% at 10 K and ~15% at 300 K and reaches a maximum of ~21.5% at 681 K, indicating that the enhancement of modulus is greater at higher temperatures. Conversely, a reverse trend is observed in the case of yield stress. The percentage improvement in yield strength (for the NC Al-(30,30) CNT compared to the NC Al without CNT) is ~34% at 10 K, ~28% at 300 K, and ~20% at 681 K. Furthermore, as the temperature increases, the modulus and yield strength decrease, but the ultimate tensile strength exhibits an opposite trend.

The stress–strain graphs indicate that the stress drops continuously beyond the UTS for NC Al without a CNT specimen at all the temperatures considered in this study. In contrast, for the NC Al-CNT nanocomposites, the stress increases slightly beyond the UTS regime. This trend is more pronounced for the composite specimens with higher CNT diameter. This behavior is attributed to the contribution of the matrix to the total stress, which is higher at lower strains due to its higher volume fraction, and the contribution of CNT to the total stress, which is higher at higher

TABLE 6.1

Mechanical Properties of NC Al and CNT-Reinforced NC Al Composites for Different Temperatures

Specimens	Young's Modulus (GPa)			Yield Strength (GPa)			Ultimate Tensile Strength (GPa)			Fracture Strain		
	10 K	300 K	681 K	10 K	300 K	681 K	10 K	300 K	681 K	10 K	300 K	681 K
NC Al without CNT	100.9	97.8	91.3	3.2	3	2.9	6.52	6.93	6.91	0.56	1.02	1.31
NC Al with (5,5) CNT	104.1	101	100	4.1	3.15	2.98	6.92	7.24	7.17	0.73	0.84	1.51
NC Al with (15,15) CNT	105.6	104	102	4.2	3.4	3.12	7.14	7.33	7.28	0.63	1.01	1.71
NC Al with (30,30) CNT	115.2	113	111	4.3	3.85	3.5	8.25	8.37	8.64	0.88	1.59	2.2

TABLE 6.2

Mechanical Properties at 300 K Calculated Using MD Simulations for NC Al and CNT-Reinforced NC Al Composites with Different Grain Sizes

Specimens	Grain Size (nm)	Ultimate Tensile Strength (GPa)	Fracture Strain
NC Al without CNT	4	8.25	1.24
NC Al with (30,30) CNT	4	8.93	1.42
NC Al without CNT	6	7.68	0.81
NC Al with (30,30) CNT	6	8.46	1.31
NC Al without CNT	9	6.93	1.02
NC Al with (30,30) CNT	9	8.37	1.59

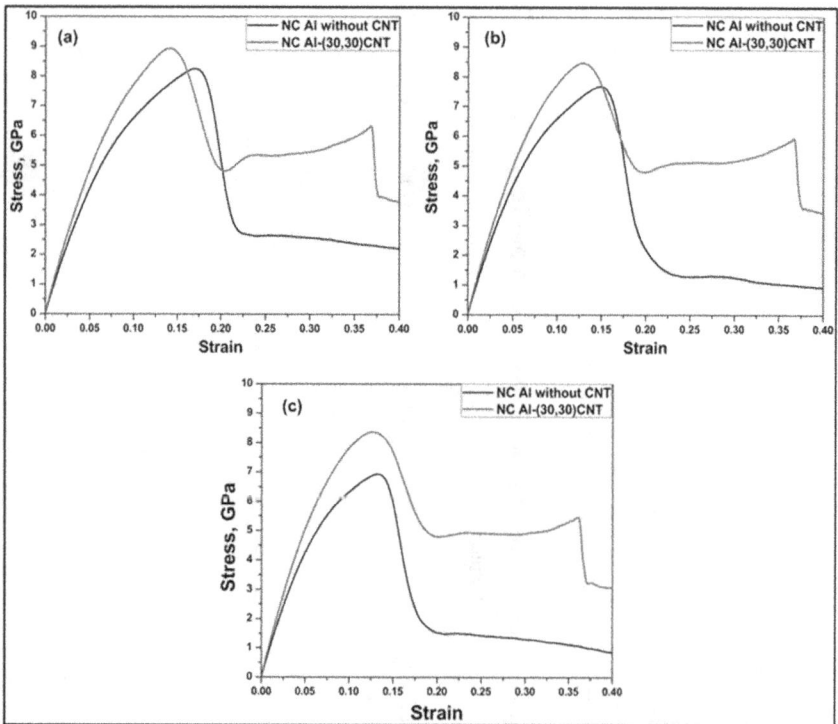

FIGURE 6.5 Stress–strain plots of NC Al without CNT and CNT-reinforced NC Al specimens at room temperature for different grain sizes: (a) 4 nm, (b) 6 nm, and (d) 9 nm.

strains due to its higher strength. In other words, the stress–strain graphs of the matrix and the CNT follow the rule of mixtures.

Table 6.2 presents the effect of grain sizes on the mechanical properties of NC Al and CNT-reinforced NC Al composites at 300 K, while the corresponding stress–strain plots are shown in Figure 6.5. The results reveal that the ultimate tensile strength is highest for the smallest grain size studied. Our findings are consistent

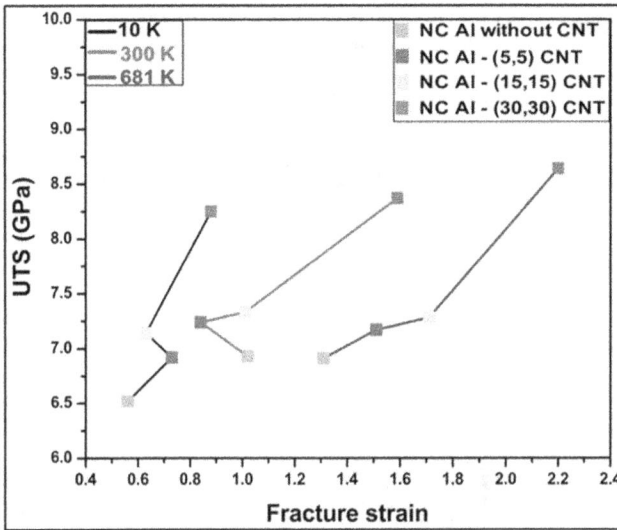

FIGURE 6.6 (a) Ultimate tensile strength vs. fracture strain plot showing NC Al without CNT and CNT-reinforced NC Al specimens at various temperatures: 10 K, 300 K, and 681 K.

TABLE 6.3
Volume Fraction of (5,5) CNT, (15,15) CNT, and (30,30) CNT

Chirality	Volume Fraction (in %)
(5,5)	0.075
(15,15)	0.224
(30,30)	0.448

with those reported in the literature [Xu and Davila, 2017; Tsai et al., 2017], which indicate that increasing grain size results in decreased mechanical properties. In many engineering applications, strength is crucial, but toughness is also a significant consideration. To this end, Figure 6.6 compares the ultimate tensile strength and fracture strain achieved by CNT reinforcement. It is apparent that the addition of (5,5) and (15,15) CNTs resulted in only minor improvements in the ultimate tensile strength and minor changes in the fracture strain. However, the addition of (30,30) CNT induced a substantial improvement in both the ultimate tensile strength and fracture strain, indicating a higher toughness at all the considered temperatures.

The CNT volume fractions investigated in this study are summarized in Table 6.3, while the properties of the resulting nanocomposites are shown in Figure 6.7. The graphs depict the mechanical properties of the materials with CNT volume fractions ranging from 0 vol. % (without CNT) to 0.4 vol. %, indicated by the colors red, green, blue, and cyan, respectively. The results show that the Young's modulus and yield strength increase with the CNT volume fraction, which is consistent with previous

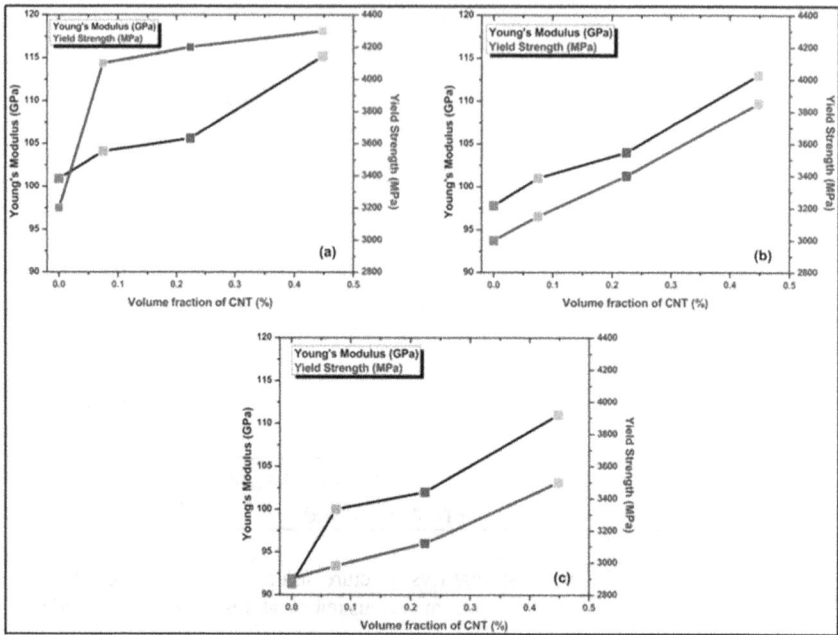

FIGURE 6.7   Mechanical properties for varying CNTs volume fraction at (a) 10 K, (b) 300 K, and (c) 681 K.

studies [Choi et al., 2011; Chunfeng et al., 2007; Liu et al., 2014; Ostovan et al., 2015]. Furthermore, an increase in the CNT volume fraction also enhances the load-bearing capacity, UTS, and fracture strain of the composite, which is also in agreement with prior literature [Kwon et al., 2009, 2010].

The stress–strain plots of the NC Al and NC Al-CNT specimens at a strain rate of 10^9 s^{-1} and 300 K are depicted in Figure 6.8. It is observed that the stress–strain plots at 10^9 s^{-1} display lower ultimate tensile strength values and a sudden decline in strength after UTS in comparison to the stress–strain plots at 10^{10} s^{-1}. Nevertheless, the mechanical properties of the different specimens examined, such as Young's modulus, yield strength, and UTS, remain unchanged.

6.4.2   DISLOCATION ANALYSIS OF NC AL AND CNT-NC AL NCS

Defects in materials often play a crucial role in determining their mechanical behavior. As shown in Figures 6.9(a–c), the dislocation density evolution during tensile deformation follows a similar trend for all cases. Initially, the dislocation density decreases as dislocations move towards the surface of the specimen, while dislocation generation is limited due to higher annihilation rates at the surface. However, as deformation progresses, the rate of dislocation generation increases, resulting in a gradual increase in dislocation density, which eventually reaches a steady state at the end of the process. The black vertical lines in the plots indicate the strain at UTS,

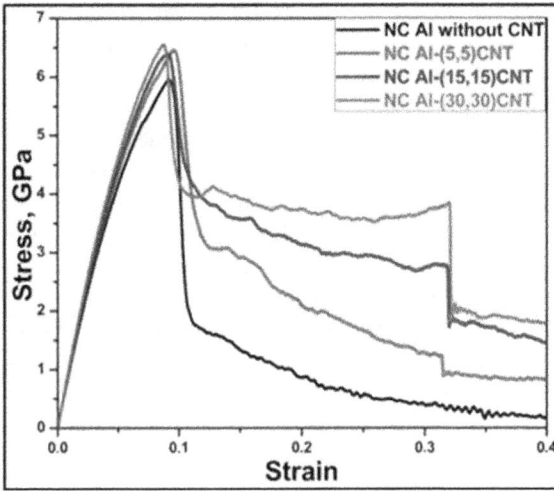

FIGURE 6.8 Stress–strain plots of the NC Al with and without CNTs at 300 K for a 10^9 s^{-1} strain rate.

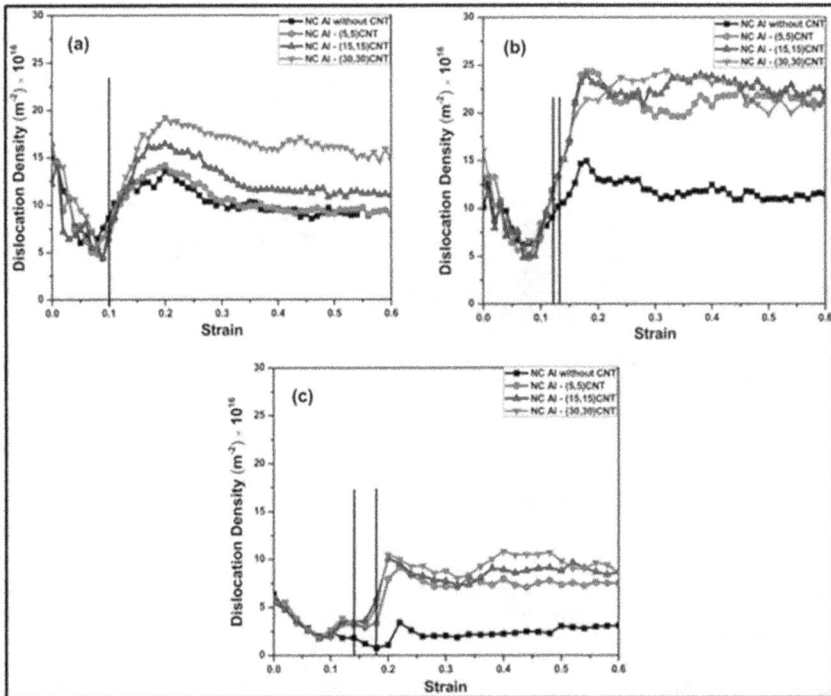

FIGURE 6.9 Plots of overall dislocation densities vs. strain for NC Al without CNT and CNT-reinforced NC Al specimens at various temperatures: (a) 10 K, (b) 300 K, and (c) 681 K.

which varies with the CNT diameter. Interestingly, the increase in dislocation density beyond the UTS point is found to be higher for NC Al composites with higher CNT diameter and at higher temperatures. The effect of temperature on dislocation density is significant. At lower temperatures, such as 10 K, dislocation density is lower than expected from strain hardening theory. This may be due to the low mobility of atoms and high strength of grain boundaries, which make dislocation generation difficult. In NC materials, dislocation generation occurs mainly at grain boundaries, where the proximity of boundaries attracts dislocations towards themselves via image dislocations [Frøseth et al., 2004; Derlet et al., 2003]. The inability of grain boundaries to produce dislocations due to their small size and proximity to each other reduces the overall dislocation density. Moreover, CNT-reinforced NC Al composites exhibit a higher dislocation density than NC Al alone, especially at higher temperatures. This can be attributed to the hindrance of dislocation movement by CNT, owing to the mismatch between the matrix and the CNT particles at the interface. Interfacial shear stresses or interactions depend on the CNT chirality and interaction perimeter. Studies by Xiang et al. and Lu et al. have shown that CNT-chirality–dependent interfacial shear stresses exist at the matrix–CNT interface. [Lu et al., 2016].

Figure 6.11 provides a detailed analysis of the dislocation density, specifically for four types of dislocations: perfect, Shockley partial, Frank partial, and stair-rod. These dislocations are crucial for understanding the mechanical behavior of materials. The data presented in Figure 6.11 corresponds to a temperature of 10 K. Upon comparing Figure 6.11(a) with 6.11(b), 6.11(c), and 6.11(d), it becomes evident that

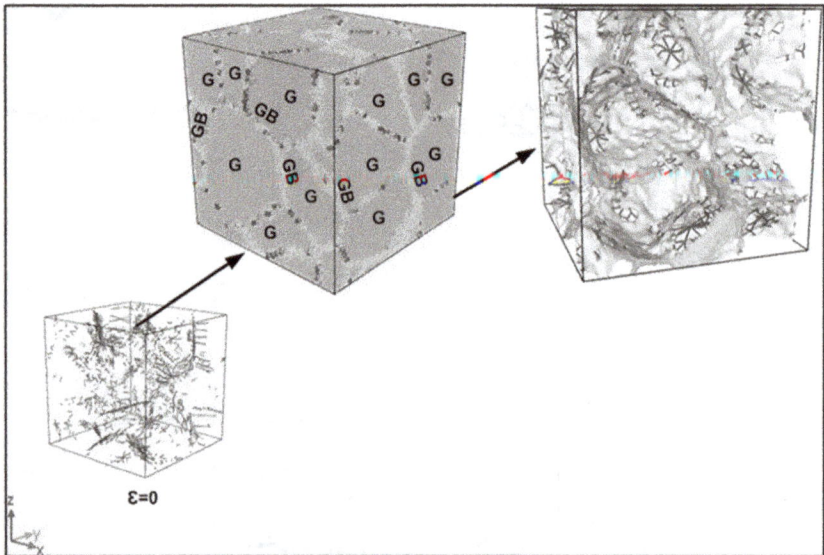

FIGURE 6.10   Atomic snapshots using dislocation analysis of (30,30) CNT-reinforced NC Al nanocomposite specimen at initial stage (before loading) $\varepsilon = 0$. (Inset: common neighbor analysis showing the dislocations (various colors) in the grain boundary; grains are in green, grain boundaries are in white.)

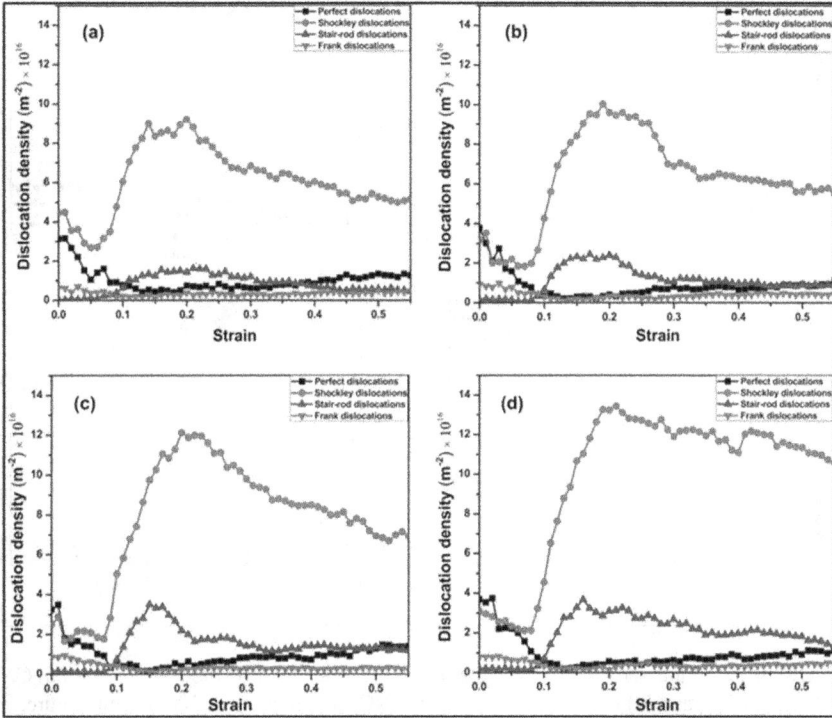

FIGURE 6.11 Plots showing types of dislocations for (a) NC Al without CNT, (b) (5,5) CNT, (c) (15,15) CNT, and (d) (30,30) CNT-reinforced NC Al specimens at 10 K temperature.

adding CNT to NC Al promotes the formation of Shockley and stair-rod dislocations. During the initial stages of deformation, the density of both Shockley partials and stair-rod dislocations remains mostly unchanged. However, there is a sudden increase in their densities at a specific strain, which eventually decreases at around 17–22% strain, where a drop in these densities is observed. Conversely, there is not much variation observed in the dislocation density for perfect and Frank partial dislocations as a function of CNT diameter. Additionally, the dislocation densities decrease in the initial stages of deformation and remain constant thereafter.

At 300 K, the variations in dislocation density are similar to those observed at 10 K. However, the density of perfect dislocations increases slightly in the later stages of deformation (as depicted in Figure 6.12). The presence of stair-rod and Frank dislocations is not observed at 681 K (see Figure 6.13). This indicates that deformation primarily occurs through Shockley partials and, to a lesser extent, perfect dislocations. This is due to the high strain rates and stresses that are achieved, which exceed the critical stress required for initiating partial dislocation activity. The critical stress (τ) is dependent on the intrinsic stacking fault energy (γ) and the Burgers vector (**b**). For the stacking fault to remain stable, the work must exceed γ, that is, $\tau > \gamma/\mathbf{b}$. As the stresses in this study are well above ~1 GPa, partial dislocations (manifested as stacking faults) are dominant, and most of the perfect dislocations are expected to

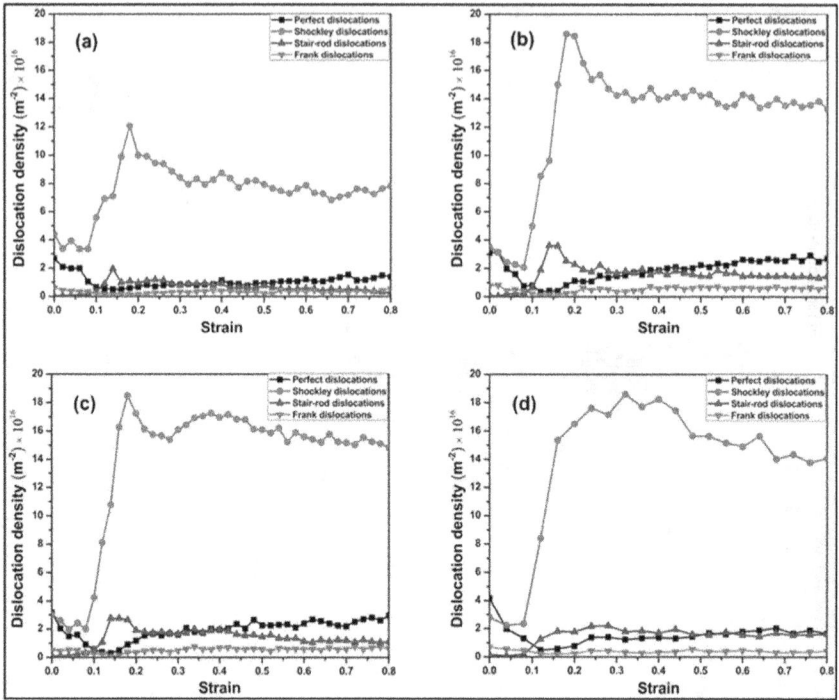

FIGURE 6.12 Plots showing types of dislocations for (a) NC Al without CNT, (b) (5,5) CNT, (c) (15,15) CNT, and (d) (30,30) CNT-reinforced NC Al specimens at 300 K temperature.

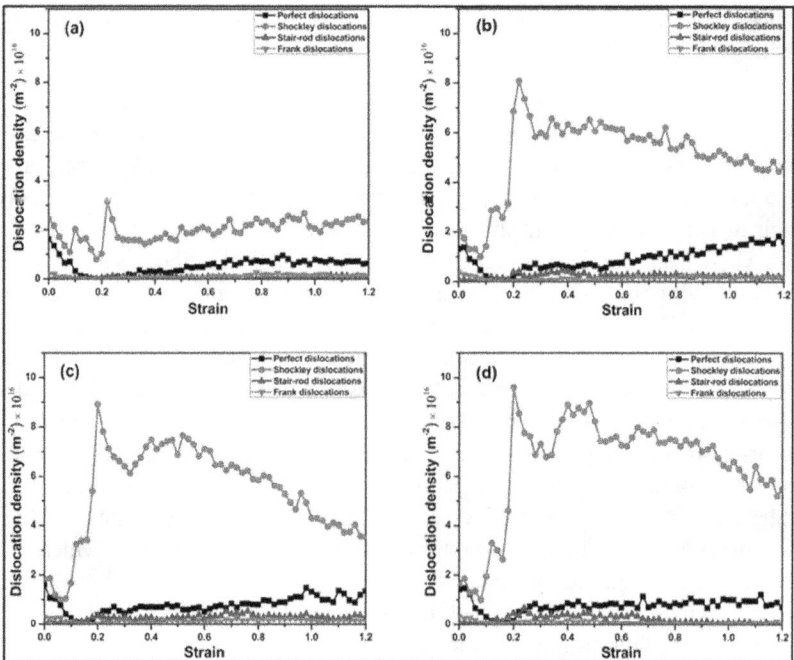

FIGURE 6.13 Plots showing types of dislocations for (a) NC Al without CNT, (b) (5,5) CNT, (c) (15,15) CNT, and (d) (30,30) CNT-reinforced NC Al specimens at 681 K temperature.

have dissociated into partials, which is in line with existing literature [Chen et al., 2003; Hirth and Lothe, 1992].

6.4.3   STRUCTURAL ANALYSIS OF NC AL AND CNT-NC AL NCS

Figure 6.14 displays atomic snapshots of (30,30) CNT-reinforced NC Al specimens during tensile deformation process in slice vision at 10 K temperature. In Figure 6.14, atoms are colored based on the shear strain, and the figures demonstrate the various features influencing the deformation in NC Al with (30,30) CNT. At 14% strain, void generation at the interface between the grain boundary of the matrix and CNT initiates failure of the composite. CNA images also suggest the formation and propagation of stacking faults, and dislocation entanglement and dislocation multiple junctions are formed. The morphology of the CNT shows no cracks, indicating matrix cracking as the mechanism of failure of the composite due to the low ductility of metals at such temperatures. The cracks propagate along the grain boundaries via void amalgamation. In Figure 6.14(c), the CNT fractures at 38% strain. In contrast, the atomic snapshots of NC Al with (30,30) CNT at 300 K in Figure 6.15 demonstrate a similar deformation behavior to 10 K, except for high strains. Failure is initiated by cracking in nanocrystalline materials followed by carbon nanotube at room temperature.

In Figure 6.16, we can see the atomic snapshots of the various features that influence the deformation in NC Al with (30,30) CNT at a higher temperature of 681 K. At 22% strain, voids begin to form, and a comparison with Figure 6.15(a) suggests that the grain boundary has thickened at higher temperatures. Additionally, the shear strain distribution is non-uniform even at lower strain values. This indicates that the

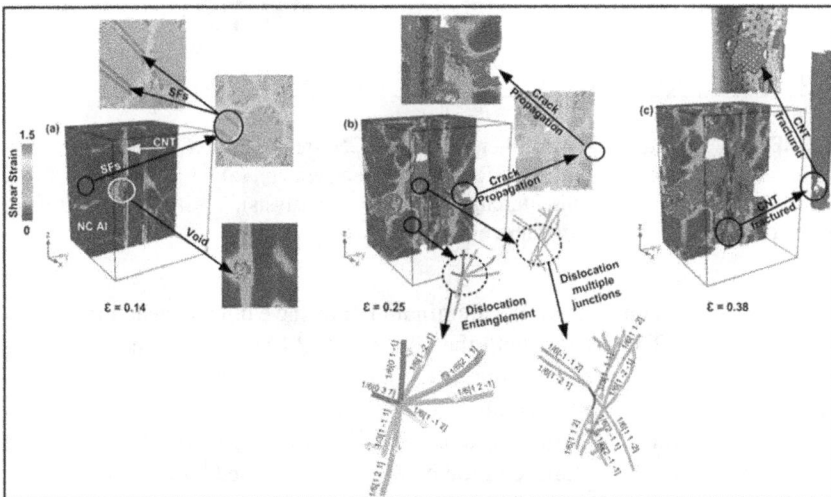

FIGURE 6.14 The atomic strain snapshots in slice view of NC Al specimens reinforced with (30,30) CNTs during the tensile deformation process at 10 K. These snapshots reveal the key stages of deformation: (a) the formation of voids and stacking faults (inset via common neighbor analysis), (b) crack propagation and dislocation representation, and (c) fracture of the CNT.

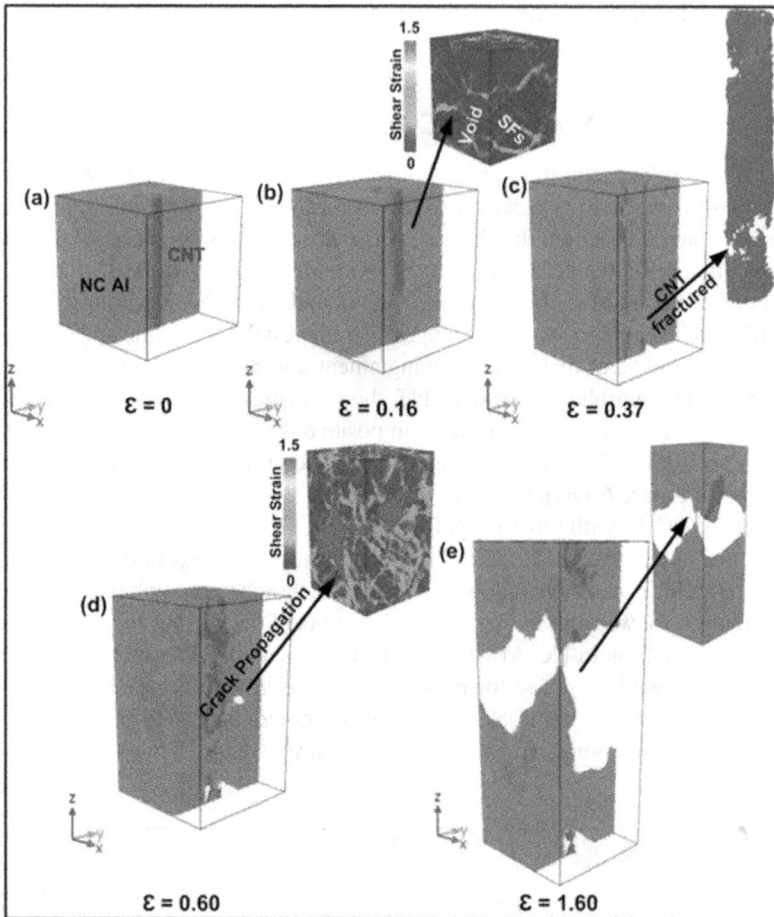

FIGURE 6.15 Atomic strain snapshots of (30,30) CNT-reinforced NC Al specimens during tensile deformation process in slice vision at 300 K temperature. (a) Initial specimen, (b) void and stacking fault formation (inset through atomic strain analysis), (c) CNT fracture, (d) crack propagation, and (e) complete fracture of nanocomposite specimen.

regions of CNT in contact with the aluminum lattice have higher shear strains, while the regions of CNT in contact with the GBs of NC Al have lower shear strains. At a strain of 36%, CNT fracture is initiated, and there is significant inhomogeneity in the strain distribution, with fracture occurring in regions with higher strains. The snapshot at a strain of 65% in Figure 6.16(c) shows that the matrix begins to crack. Unlike the 10 K case, the failure of the composite is initiated by fiber breakage and continued by matrix cracking. This is expected at higher temperatures due to the matrix's increased ductility and the CNT reinforcement's relatively brittle nature. In the vicinity of the crack, Shockley partials, stair-rod, and perfect dislocations are seen, and a majority of the dislocations happen to be either edge or mixed types of dislocations. This is because screw dislocations have high mobility, and they move

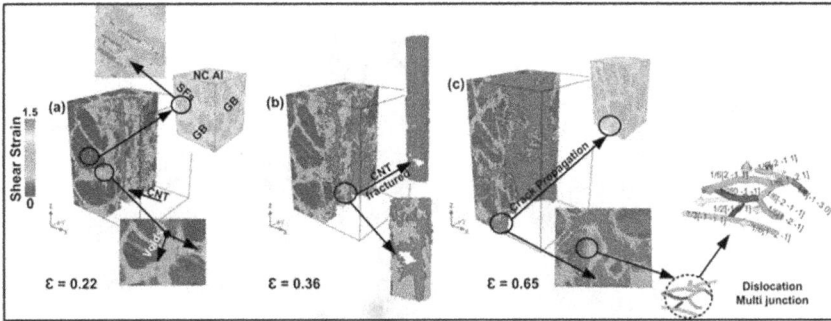

FIGURE 6.16 During tensile deformation process, the snapshots shows the following: atomic strain snapshots of (30,30) CNT-reinforced NC Al specimens in slice vision at 681 K temperature. (a) CNA snapshots of the void formation and stacking fault formation, (b) CNT fracture, and (c) crack propagation and representation of dislocations.

quickly in response to the applied load and get annihilated at the grain boundaries or the surfaces. Overall, the sequence of events: void formation, crack propagation at the grain boundary, and CNT fracture is observed during the tensile deformation of all the CNT-reinforced specimens, as seen in Figure 6.16.

6.4.4 FRACTURE BEHAVIOR OF CARBON NANOTUBES

In Figure 6.17, the mechanism of CNT failure for (5,5) CNT at a low temperature of 10 K is presented. This failure mechanism is applicable to all diameters. The "fiber" appearance indicates that the bonds are broken by shear failure, which means that bond rupture occurs. The disrupted order of atoms from where the failure has been initiated is depicted in Figure 6.17(e). Interestingly, the point of rupture of the CNT coincides with the point where voids developed in the Al matrix. Additionally, the formation of Stone–Wales defects is another mechanism that can lead to CNT failure. These defects can weaken the CNTs and make them more susceptible to failure, as noted in the literature [Choi et al., 2016]. The CNT morphology depicted in Figure 6.17(f) highlights the regions where the rings are distorted from hexagonal shapes to rings with five and seven carbon atoms, which are Stone–Wales defects. These defects can cause distortion in the bonds and bond angles, thus becoming sites of failure initiation and rapid propagation, as previously reported by Lu and Bhattacharya [2005]. The fracture mechanism observed for (15,15) CNT and (30,30) CNT is similar to the mechanism observed for (5,5) CNT.

6.4.5 VACANCY ANALYSIS OF NC AL AND CNT-NC AL NCs

While dislocations and shear strain distribution provide valuable information about material deformation behavior, vacancies also play a critical role, especially at high temperatures, due to their increased concentration and mobility. In nanocrystalline materials, vacancies can diffuse easily through grain boundaries, making vacancy motion an essential factor to consider. Figure 6.18 illustrates the variation of the total

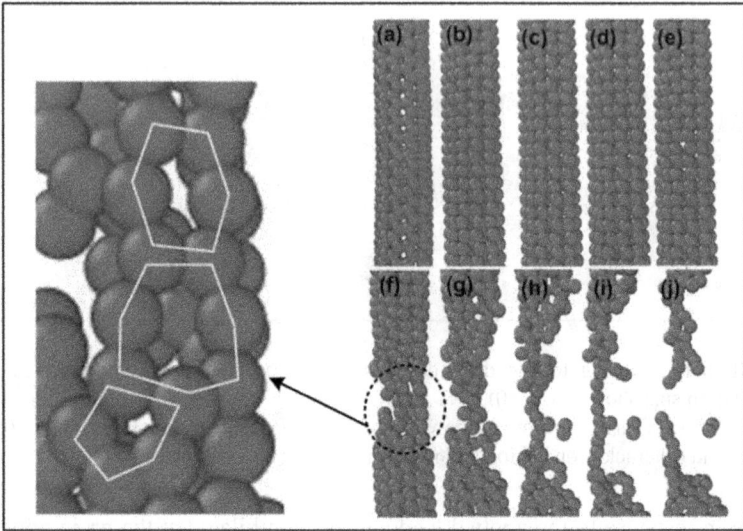

FIGURE 6.17 Atomic snapshot representations of (5,5) CNT fracture path for 10 K during tensile deformation process.

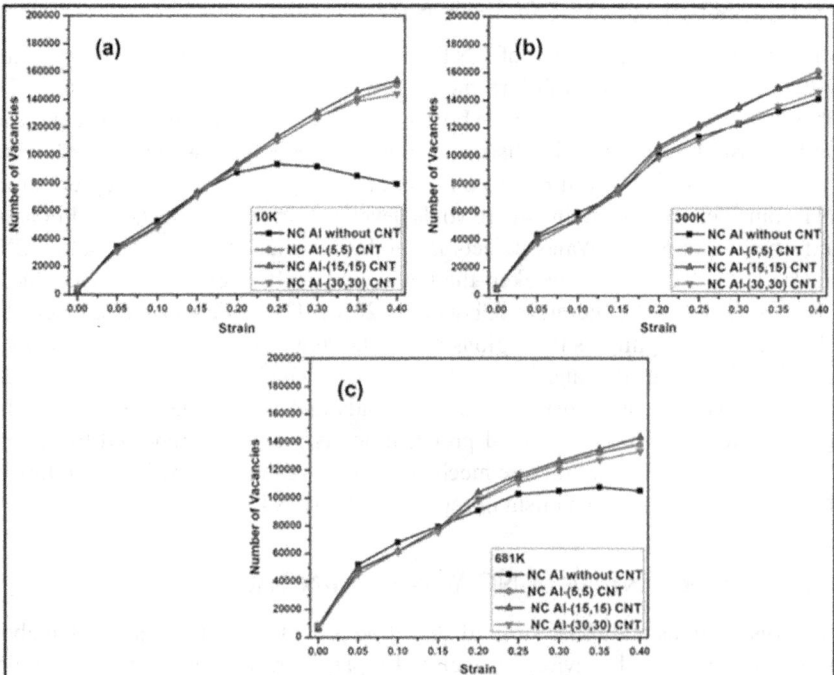

FIGURE 6.18 Plots of the number of vacancies vs. strain for NC Al without CNT and CNT-reinforced specimens at various temperatures: (a) 10 K, (b) 300 K, and (c) 681 K.

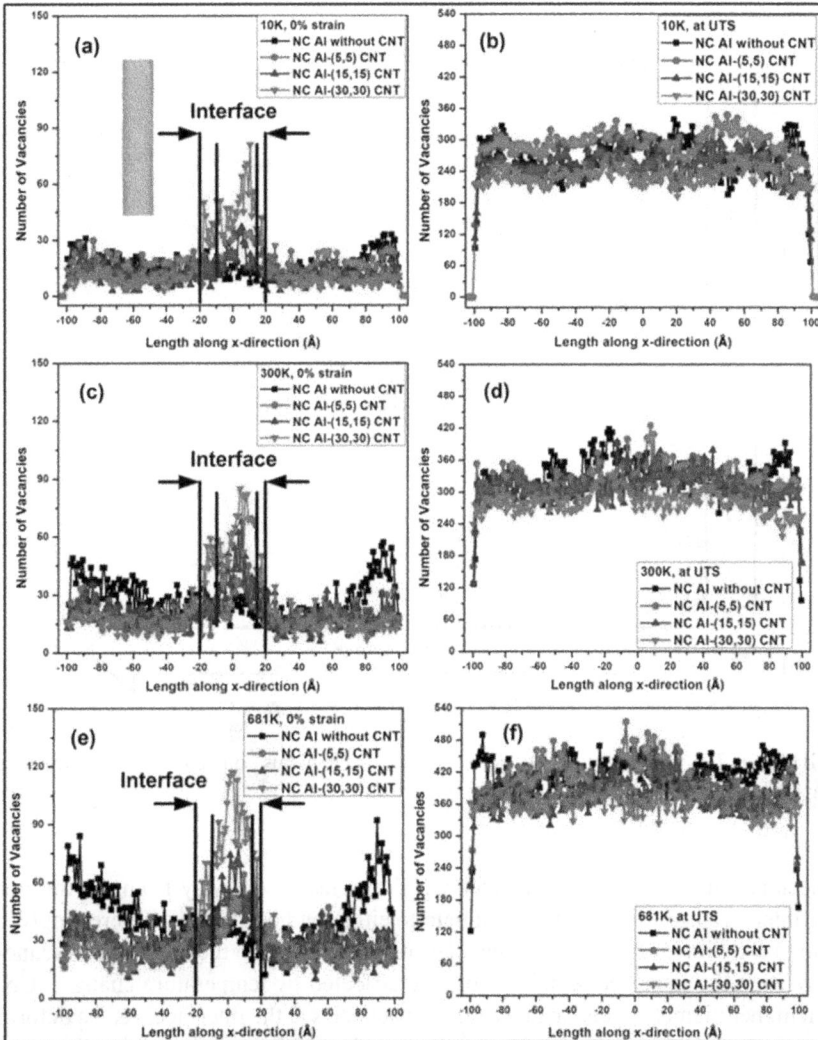

FIGURE 6.19 Plots showing number of vacancies vs. length along the *x*-direction for NC Al without CNT and CNT-reinforced NC Al specimens at various temperatures (a) 10 K—0% strain, (b) 10 K—UTS, (c) 300 K—0% strain, (d) 300 K—UTS, (e) 681 K—0% strain, and (f) 681 K—UTS.

number of vacancies with strain for our specimens at different temperatures—10 K, 300 K, and 681 K. All three figures show that vacancies are produced as the deformation progresses. At both 10 K and 681 K, the NC Al displays a reduced number of vacancies at later stages of deformation. The effect of CNT diameter on the number of vacancies is insignificant across all temperatures investigated. However, Figure 6.19 provides better insight into the effect of CNT diameter on vacancy concentration through its spatial distribution along the *x*-direction. It is evident that the vacancy

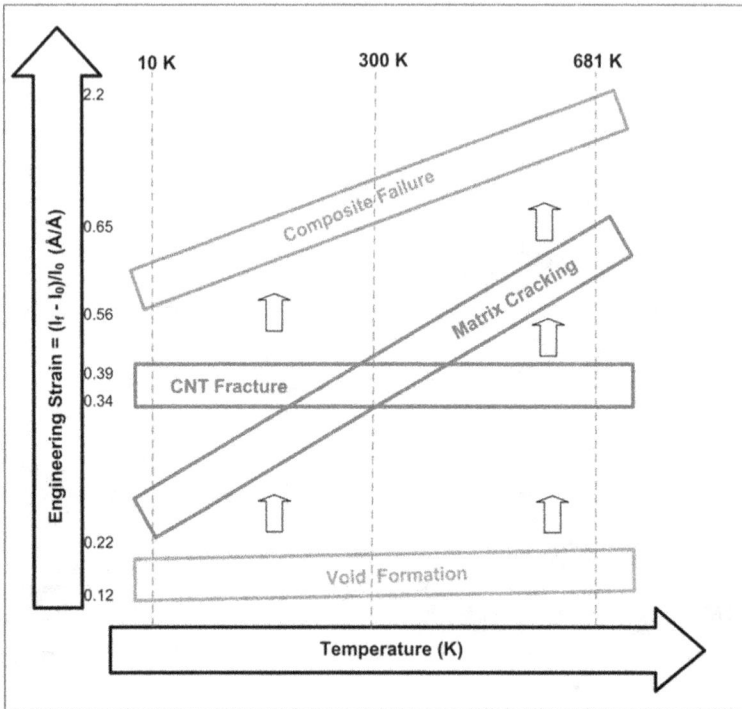

FIGURE 6.20 Dependence of deformation mechanisms on engineering strain and temperature (l_f—final length and l_0—initial length).

concentration is higher at the matrix–CNT interface than away from the interface. The interface here refers to the region up to either end of the CNT. Interestingly, the ratio of the average number of vacancies at the interface to the number of vacancies away from the interface remains mostly unaffected by temperature changes. CNTs with higher diameters tend to attract more vacancies at the interface. As the deformation progresses, the vacancy pileup at the interface is destroyed, and the distribution of vacancies becomes uniform across the specimen, likely via grain boundary diffusion of vacancies at low temperatures (such as 10 K) and lattice diffusion at high temperatures (such as 681 K), as previously reported in literature [Wilde and Divinski, 2019]. Additional vacancies are also produced during deformation.

In this study, various deformation mechanisms were observed and analyzed in the nanocrystalline aluminum matrix reinforced with carbon nanotubes at different temperatures. The results show that at all investigated temperatures, void nucleation occurs at grain boundaries of the matrix and grain boundary–CNT interfaces, with the strain range of 0.12 to 0.22. Matrix cracking occurs through void amalgamation along the grain boundary, with the initiation of cracking observed at low strains for low temperatures (~10 K) and high strains for high temperatures (~681 K). Shockley partials are the predominant driving deformation for NC Al without CNT and CNT-incorporated NC Al specimens. The dislocation density is higher in NC

Al specimens with CNTs. CNT fracture due to Stone–Wales defect occurs independently of temperature in the strain range of 0.34 to 0.39. The composites tend to fail in the following order: first, the NC Al matrix material fractures, followed by the CNT due to low ductility at low temperatures, while at high temperatures, the CNT fractures first, followed by the NC Al matrix material. The results suggest that CNT incorporation has a significant effect on the deformation behavior of NC Al, and the observed mechanisms can provide insight into the design of future composites with improved mechanical properties.

A study using molecular dynamics simulation investigated the impact of incorporating CNTs on the mechanical behavior of random grain boundaries in NC Al and NC Al–reinforced CNT nanocomposites at different temperatures. The following conclusions were drawn from the results:

- Grain boundaries of the matrix and grain boundary–CNT interfaces are potential sites for void nucleation for all the investigated temperatures. (30,30) CNT is found to be the most effective in enhancing the strength and delaying fracture in NC Al compared to other nanocomposite specimens.
- CNT volume fraction increases Young's modulus, yield strength, load-bearing capacity, UTS, and fracture strain of the composite.
- CNTs may fail due to the rupture of bonds by shear failure or the formation of Stone–Wales defects, which decrease their strength.
- In CNT-reinforced NC Al nanocomposites, matrix cracking initiates failure at low temperatures, while fiber breakage initiates failure at high temperatures.
- Vacancy concentration is higher for higher CNT diameters at the interface due to the larger surface area. These findings provide insights into the underlying mechanisms responsible for improving the mechanical properties of NC metals reinforced with CNTs.

6.5 EXAMPLE LAMMPS INPUT CODE

```
#----------------------------INITIALIZATION----------------------------
units           metal
echo            both
dimension       3
boundary        p    p       p
atom_style      atomic
#--------------------------ATOM DEFINITION----------------------------
read_data       NCAl-CNT(30,30)
mass 1 26.981
mass 2 12.000
#--------------------------FORCE FIELDS----------------------------
pair_style hybrid lj/cut 10.2 eam/alloy airebo 2.5 1 0
pair_coeff * * eam/alloy Al99.eam.alloy Al NULL
pair_coeff * * airebo CH.airebo NULL C
pair_coeff 1 2 lj/cut 0.038 2.96
#--------------------------SETTINGS----------------------------
compute csym all centro/atom fcc
```

```
compute peratom all pe/atom
# ENERGY MINIMIZATION
dump            1 all custom 50 Al_CNT_en_min*.lammpstrj id type x y z
minimize        1.0e-4 1.0e-6 1000 1000
undump          1
# EQUILIBRATION
reset_timestep 0
timestep 0.001
velocity all create 10 12345 mom yes rot no
fix 1 all npt temp 10 10 0.01 iso 0.0 0.0 0.1
#Set thermo output
thermo 1000
thermo_style custom step lx ly lz press pxx pyy pzz pe temp
# Run for at least 10 picosecond (assuming 1 fs timestep)
run 20000
unfix 1
# Store final cell length for strain calculations
variable tmp equal "lz"
variable L0 equal ${tmp}
print "Initial Length, L0: ${L0}"
# DEFORMATION
reset_timestep 0
fix             1 all nvt temp 10 10 0.01
variable srate equal 1.0e10
variable srate1 equal "v_srate/1.0e12"
fix             2 all deform 1 z erate ${srate1} units box
# Output strain and stress info to file
# for units metal, pressure is in [bars] = 100 [kPa] = 1/10000 [GPa]
# p2, p3, p4 are in GPa
variable strain equal "(lz—v_L0)/v_L0"
variable p1 equal "v_strain"
variable p2 equal "-pxx/10000"
variable p3 equal "-pyy/10000"
variable p4 equal "-pzz/10000"
fix def1 all print 100 "${p1} ${p2} ${p3} ${p4}" file NcAlCNT_Tensile_100.def1.txt
# Use cfg for AtomEye
dump            1 all custom 1000 Nc_Al_CNT.tensile_*.lammpstrj id type x y z
# Display thermo
thermo          1000
thermo_style    custom step v_strain temp v_p2 v_p3 v_p4 ke pe press
run             2000000
# SIMULATION DONE
print "All done"
```

7 Material Processing Using MD Simulation
Nanoscale Rolling Process

Material processing techniques are critical in designing and engineering materials with desirable properties. With the advent of computational simulation tools such as molecular dynamics simulation, it has become possible to investigate the underlying physics behind various material processing techniques at the atomistic level. One such technique is the nanoscale rolling process, which has been shown to have a significant influence on the structural properties of nanomaterials. The nanoscale rolling process is particularly interesting, as it allows for the tuning of the crystallographic orientations of metallic systems. This, in turn, can have a significant impact on the mechanical properties of the material. However, understanding the deformation mechanism at the nanoscale level has been challenging due to constraints in instrument setup and operational cost. By using MD simulations, it is possible to gain insights into the atomistic mechanisms during the nano-rolling process of crystalline, amorphous, and nanolaminate metallic systems. This chapter aims to provide a thorough investigation of the nanoscale processing of various metallic systems using MD simulations. The original contribution of this chapter is to offer insights into the atomistic mechanisms during the nano-rolling process of different metallic systems, including single-crystal, nanocrystalline, and metallic glass systems. The chapter will also investigate the effects of various processing parameters on the rolling behavior of these metallic systems.

7.1 MATERIAL PROCESSING OF NANOSTRUCTURED MATERIALS

Nano-processing techniques are becoming the most significant to design, develop, and tailor nanoscale materials and can cause constructive alterations in the structure and properties of nanomaterials to suit the intended applications in electronics, medicine, energy, and other fields [Yoshino et al., 2009; Xu et al., 2017; Ji et al., 2013; Deng et al., 2015; Goswami et al., 2018; Murashkin et al., 2016; Taskaev et al., 2015]. In most cases, nanoscale metals and alloys are produced in the form of thin films by processing techniques such as chemical- or electro-deposition [Li and Ebrahimi, 2003] and sputtering [Mitra et al., 2001]. This section will discuss some of the most common nanoscale material processing techniques, including chemical vapor deposition, molecular beam epitaxy, electrospinning, and rolling. One of the most widely used nanoscale material processing techniques is chemical vapor deposition. CVD is a process that involves the deposition of a thin film of material onto a substrate through the reaction of vapor-phase chemicals. The substrate is heated to a high

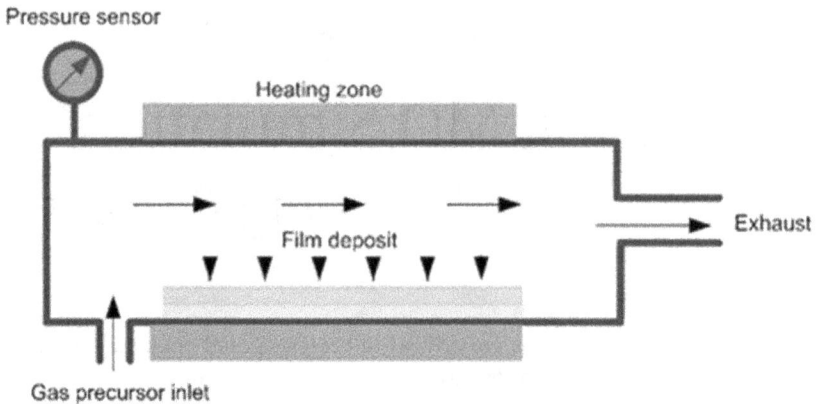

FIGURE 7.1 Schematic diagram of a chemical vapor deposition system.

temperature, and a gas containing the desired material is introduced into the reaction chamber. The gas reacts on the surface of the substrate, forming a thin film of the material. CVD is a versatile technique that can be used to produce a wide range of materials, including metals, semiconductors, and ceramics. It is widely used in the electronics industry to produce thin films for integrated circuits, sensors, and other devices. CVD is also used in the production of carbon nanotubes and graphene, two materials that have great potential for a wide range of applications.

Another widely used nanoscale material processing technique is molecular beam epitaxy (MBE). MBE is a process that involves the deposition of thin films of material onto a substrate through the controlled deposition of individual atoms or molecules. In MBE, a beam of atoms or molecules is directed at the surface of the substrate, where they bond to form a thin film. MBE is a highly precise technique that is widely used in the production of semiconductor devices, including transistors and laser diodes. It is also used in the production of high-performance solar cells, which require the precise control of material properties on the nanoscale.

Electrospinning is another important nanoscale material processing technique. Electrospinning is a process that involves the production of nanofibers through the application of an electric field to a polymer solution. In electrospinning, a high voltage is applied to a polymer solution, causing the solution to form a jet that is then spun into nanofibers. Electrospinning is a versatile technique that can be used to produce nanofibers with a wide range of properties, including size, shape, and chemical composition. It is widely used in the production of biomaterials, including tissue engineering scaffolds and drug delivery systems. It is also used in the production of nanofiber-based sensors and electronic devices.

In conclusion, nanoscale material processing techniques are essential for the development of a wide range of applications. Chemical vapor deposition, molecular beam epitaxy, and electrospinning are just a few of the many techniques that are used to manipulate and fabricate materials on the nanoscale. As the field of nanotechnology continues to grow, it is likely that new and more advanced nanoscale material

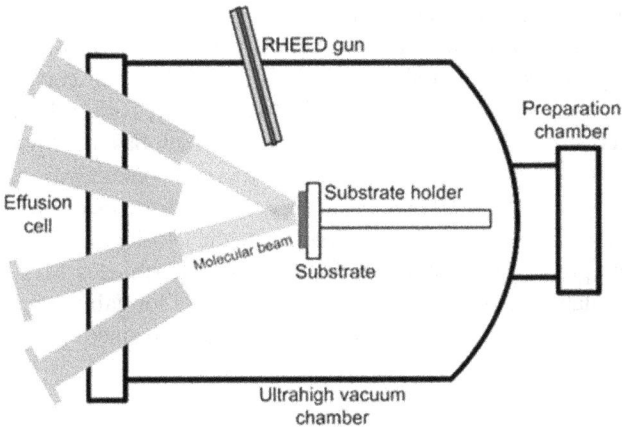

FIGURE 7.2 Schematic diagram of molecular beam epitaxy deposition.

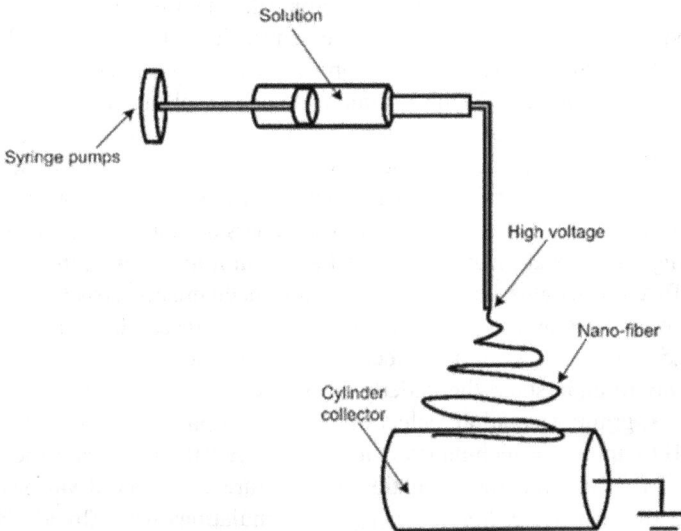

FIGURE 7.3 Schematic diagram of electrospinning material processing technique.

processing techniques will be developed, further expanding the range of potential applications for these technologies. One such potential application is controlling the deformation behavior, grain orientation, and size at the nanoscale to enhance directional properties of the metallic system. It is known that preferred grain orientations or textures can substantially influence the structural and mechanical properties of nanoscale metallic systems and subsequently define product efficiency [McCabe et al., 2014; Probst et al., 2018; Suwas and Ray, 2014]. Metallic systems can be effectively processed using various techniques to control their mechanical properties and texture. One such technique is the rolling process, which has been widely studied due to its significant influence on the plastic deformation behavior of metals and the control it has over the final product's mechanical properties [Lu et al., 2016; Odnobokova et al., 2015]. The texture of nanostructured materials is also influenced by the rolling process, which affects essential properties like corrosion behavior, elastic properties, and wave propagation [Suwas and Ray, 2014]. The crystal texture in metal sheets can be controlled and optimized, which has a considerable impact on properties such as irradiation resistance and thermal conductivity, as they depend on grain orientation. Therefore, understanding the atomistic mechanisms during the rolling process of metallic systems is crucial for the development of new materials with desirable properties. Computational simulation techniques such as molecular dynamics simulations have emerged as a powerful tool for gaining insights into the underlying physics of the nanoscale rolling process [Miyamoto et al., 2009; Yuan et al., 2013].

7.2 NANOSCALE ROLLING PROCESS

The rolling process has been widely used to deform bulk nanostructured metallic systems and control their mechanical properties, texture, and plastic deformation behavior. However, its feasibility at the nanoscale is still uncertain. To address this issue, researchers have recently proposed a processing technique known as the "nano-rolling process". This technique involves rolling nanoscaled metallic systems into sheets to study the underlying deformation mechanisms and texture evolution. While the experimental setup for this technique is still undetermined, advancements in material processing technology have made it a promising addition to the traditional rolling process. In fact, researchers have already developed nano-roll forming methods at the nanoscale that share similarities with the nano-rolling process. To gain insights into the deformation mechanisms, structural evolution, and texture analysis during the nano-rolling process, researchers can use atomistic simulation methods such as molecular dynamics. MD simulation provides an efficient way to understand the underlying physics behind nano-level deformation mechanisms, grain orientation, defect evolution, and phase transformation. In this regard, MD simulation techniques can be employed to investigate the deformation behavior, structural transformation, and texture evolution of various metallic systems at the nanoscale. Studies using MD simulations have already been conducted on this topic and have shown promising results [Reddy and Pal, 2018b, 2020; Scudino and Şopu, 2018].

7.2.1 NANO-ROLLING PROCESS OF FCC METALS

In this section, we explore the use of molecular dynamics simulations to investigate the deformation mechanisms and structural evolution of face-centered crystal specimens during the nano-rolling process. Our primary objective is to analyze the effect of crystallographic orientation and stacking fault energy (SFE) on the nano-rolling behavior of Al and Cu—commonly used representative metallic systems. Through LAMMPS, we provide readers with an idea of the simulation procedure and various analyses that can be performed using MD simulations. We start by examining the impact of SFE on the atomic-level structural changes that occur during the nano-rolling process. We consider Al and Cu specimens with varying SFE values and analyze the effect of SFE on structural evolution and deformation behavior. Specifically, we investigate the formation and destruction of stacking fault tetrahedra (SFT), which are fourfold coordinated defects formed by the partial dissociation of dislocations into stacking faults. Besides SFE, we also analyze the effect of crystal-lographic orientation on the mechanical and structural properties during the nano-rolling process. The orientation of the crystal planes relative to the rolling direction (RD) plays a crucial role in the deformation behavior and structural evolution.

First, we present a molecular dynamics study on the nano-rolling process of single-crystal (SC) aluminum specimens. The study involved setting up rigid rotating rollers to induce a 20% thickness reduction in all SC specimens during a single pass. Initially, the specimens moved at a fixed velocity of 0.95 Å ps^{-1} before passing through the rollers, where their velocity increased. To analyze the deviation in the final velocity of the specimen, we calculated the average specimen velocity after passing through the rollers.

In Figure 7.2, the distance vs. time plot during the nano-rolling process for a single-crystal aluminum specimen is presented, and the analysis demonstrates that the average velocity of the specimen initially increased to approximately 4.2 Å ps^{-1}

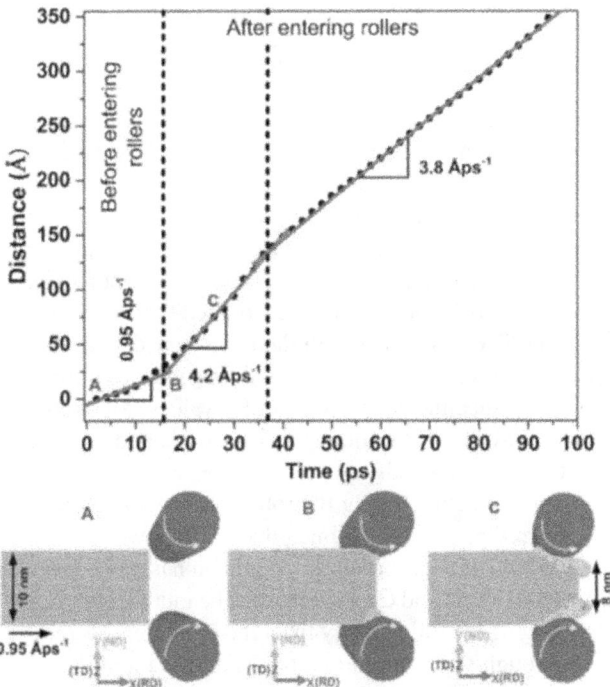

FIGURE 7.4 (a) Distance vs. time plot to calculate the rolling velocity before and after the specimen passes into the rollers. Insets (b), (c), and (d) show atomic snapshots at specific positions of the specimen.

after passing through the rollers. This increase in velocity was primarily caused by the material overflow in the rolling direction. To ensure the accuracy of the findings, the analysis can also be compared with a fundamental mathematical expression derived from the work of Dieter and Bacon [1986], which demonstrates the relationship between the initial and final velocities of the specimen after passing through the rollers.

$$h_0 V_0 w_0 = h_1 V_1 w_1 \qquad (7.1)$$

where h_0, h_1 indicate the original and final height; V_0, V_1 represent initial and final velocity; and w_0, w_1 indicate the initial and final width of the specimen, respectively. During the rolling process, it is conventionally assumed that the width of the specimen remains constant while the thickness reduces. However, for the current simulation study, periodic boundary conditions have been implemented along the z-direction to eliminate any change in the width of the specimen. Consequently, Equation (7.1) has been simplified to Equation (7.2), which demonstrates the relationship between the initial and final height and velocity of the specimen exclusively. Equation (7.2) states that the product of the initial and final height of the specimen is directly proportional to the product of its initial and final velocity. This equation is a valuable tool for analyzing the specimen's behavior during the rolling process in the simulation study.

$$\frac{h_0}{h_1} = \frac{V_1}{V_0} \qquad (7.2)$$

To study the dynamic deformation behavior during the nanoscale rolling process, MD simulations were used to analyze internal strains, defect generation, and structural alterations. The atomic strain snapshots during the room temperature nano-rolling process of Al and Cu specimens were analyzed and are presented in Figure 7.5. The analysis revealed that slip occurred along the close-packed planes of the specimens as they passed through the rollers, leading to the formation of slip bands. The extent of slip band formation was found to be directly related to the stacking fault energy of the specimen. Lower SFE values led to limited recovery during the rolling process, restricting dislocation movement and accommodating plastic strain through slip band formation. Conversely, specimens with higher SFE values exhibited a larger extent of dynamic recovery, allowing dislocations to easily glide and cross-slip, leading to a lower volume fraction of slip bands.

The formation of slip bands during the rolling process is more prominent in the Cu specimen compared to the Al specimen due to the lower SFE of Cu. The interaction of slip bands, which occurs after the specimen has passed through the rollers, has been observed in both Al and Cu specimens. The mutual interaction of slip bands helps to dissipate localized strain energy from the specimen, thereby preventing catastrophic failure through slip band formation. The formation of slip bands has been found to impede the plastic deformation of the Al specimen, whereas metallic systems with high SFE facilitate cross-slip. Slip band interaction has been identified as an active mechanism for increasing the plasticity of various material systems, as it helps to dissipate localized strain energy and prevent catastrophic failure. These

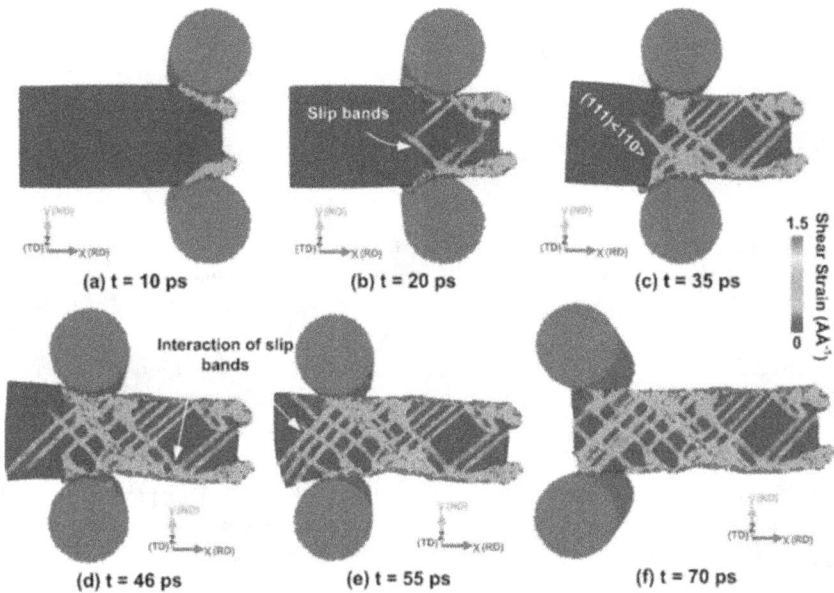

FIGURE 7.5 Atomic strain snapshots of the cube-oriented Al specimen during the room-temperature nano-rolling process at specific time periods ranging from 10 to 70 ps. The snapshots illustrate the formation of slip bands and their interaction. Each subfigure shows the atomic strain distribution at a particular time interval, with (a) corresponding to 10 ps, (b) to 20 ps, and so on until (f) at 70 ps.

findings have been supported by experimental studies reported in the literature [Varma et al., 1996; Welsch et al., 2016]. Furthermore, the MD study not only mimics the experimental findings but also provides a dynamic evolution of the deformation behavior in a frame-by-frame form, enabling a more comprehensive understanding of the nanoscale deformation process.

7.2.2 NANO-ROLLING PROCESS OF BCC METALS

The rolling process plays a crucial role in shaping the structural properties of body-centered cubic nano-structured metallic systems. However, despite its wide application in industry, the underlying atomistic mechanism of the deformation behavior of BCC materials during rolling remains largely unclear. This section aims to shed light on this area of research by employing molecular dynamics simulations to investigate the deformation mechanism and structural evolution in single-crystal Fe during cryo- and cold-rolling processes. SC Fe has been selected as a model material owing to its extensive utilization in both industrial and scientific applications. It finds application in various domains, including ferromagnetic backing frames for transient field measurements, electronic and telecommunication devices, and plates employed in orthopedic implants. Furthermore, the mechanical and magnetic properties of SC Fe are dependent on its crystallographic orientations. Although analysis

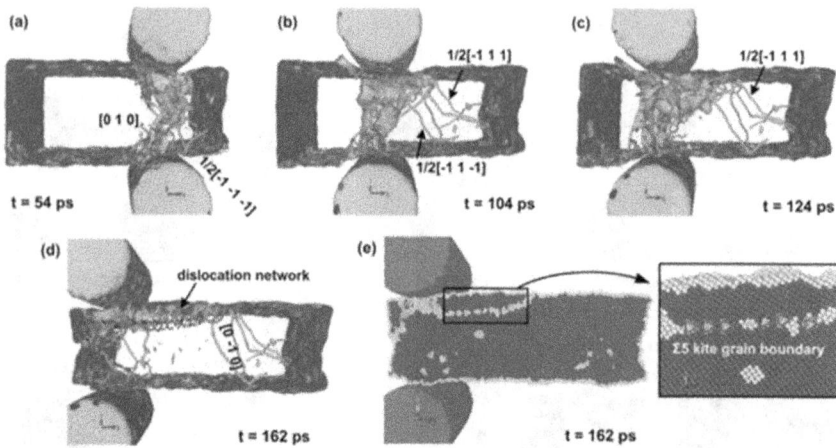

FIGURE 7.6 Atomic snapshot during the cryo-rolling (77 K temperature) process illustrating dislocation generation and grain rotation for specimen with {001}<100> orientation. Inset in (e) shows formation of Σ5 kite grain boundary structure.

such as that discussed before can be performed using BCC materials, this section focuses on virtual X-ray diffraction of unrolled and rolled specimens to show the wide scope of MD simulation in material characterization.

The SC Fe sample, possessing a {001}<100> orientation, undergoes deformation, during which sessile <001> dislocations are formed in parallel to the transverse direction. The dislocations are produced through the interaction of two 1/2<111> glissile dislocations. Furthermore, a surface mesh appears within the specimen, indicating the emergence of new surfaces, potentially resulting from phase transformation or grain boundary formation. An algorithm is employed to construct the surface mesh, which identifies the lattice points and generates a closed and oriented manifold, dividing the region into surfaces. Cryo-rolling of the sample leads to a BCC-to-FCC phase transformation, generating a new surface mesh that facilitates plastic deformation at low temperatures. However, the surface of the specimen regains a BCC phase with a different orientation, leading to the formation of misfit dislocations at the interface of two differently oriented grains, culminating in a sub-grain tilt grain boundary. During the deformation, grain rotation occurs, leading to the formation of a Σ5 kite grain boundary structure. Studies have highlighted the occurrence of such a phase transformation and grain rotation in different metallic systems during rolling [Wu et al., 2016]. The change in orientation directly impacts the diffraction patterns, which can be determined using MD simulations. The virtual XRD line profiles for unrolled, cryo-rolled, and cold-rolled specimens with {001}<001> orientation indicate a slight shift towards the right side in the peaks of (110) plane in the rolled specimens compared to the unrolled specimen. The shift is attributed to the change in crystallographic orientation caused by the formation of new grains and grain boundaries, which induces lattice strain in the sample. The formation of new grains with different orientations near the surface of the specimen leads to an increase in surface roughness in the cold-rolled specimen. These observations emphasize the crucial

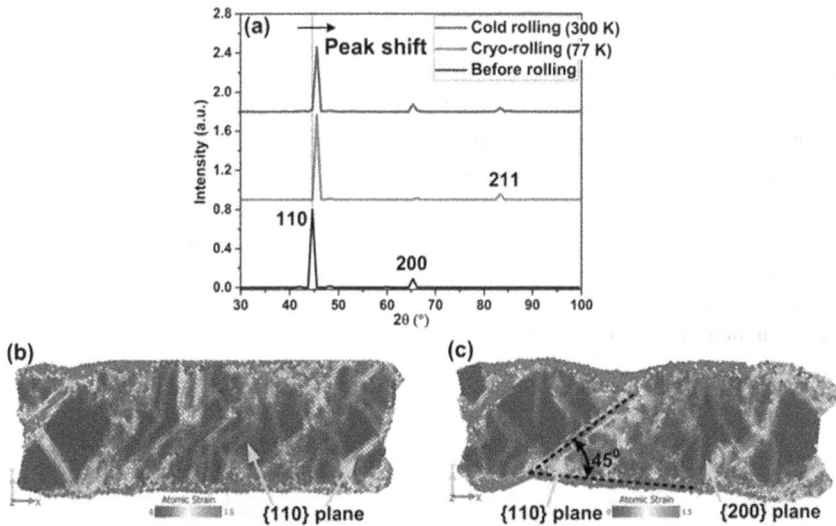

FIGURE 7.7 (a) XRD plots for specimen with {001}<100> orientation. Illustration of atomic strain in the specimen after: (b) cryo-rolling process, and (c) cold-rolling process.

role of grain orientation and boundary formation in inducing lattice strain and influencing the mechanical properties of the material.

7.2.3 NANO-ROLLING PROCESS OF HCP METALS

So far, we have discussed the deformation of cubic metallic systems during the rolling process. However, one of the most important and crucial crystal systems that has the largest influence due to the rolling process is hexagonal close-packed materials. The effect of rolling on HCP metals such as magnesium (Mg) and titanium (Ti) is different from that observed in FCC or BCC metals due to their unique crystal structures. In Mg, rolling induces a significant amount of twinning, which is the plastic deformation mechanism unique to HCP metals. Twinning in Mg is caused by the activation of basal and non-basal slip systems, which occur along the {0001} and {1012} planes, respectively. Deformation twinning helps to improve the strength and ductility of the material, but it also causes a decrease in the anisotropy of the material. Additionally, Mg has a low stacking fault energy, which makes it highly prone to cracking and fracture during rolling. Similarly, in Ti, rolling leads to a complex deformation behavior due to its HCP crystal structure. The deformation is mainly controlled by twinning and slip along the {0001} and {1011} planes, respectively. Twinning in Ti can cause a significant strengthening effect and improve the ductility of the material, but it also leads to an increase in the anisotropy of the material. However, excessive twinning can cause a decrease in the ductility of the material, leading to premature failure. Overall, the influence of rolling on HCP metals such as Mg and Ti is highly dependent on their unique crystal structure and the activation of deformation mechanisms such as twinning and slip. While rolling can improve the mechanical properties of these metals, it also requires careful consideration and control to prevent issues such as cracking, anisotropy, and premature failure. The MD simulation analysis presented in this section demonstrates the atomic-shuffle mechanism's influence on dynamic twin nucleation under compressive loads, followed by twin propagation along the grains through twin boundary disconnection steps under shear. Additionally, the nano-rolling simulation code facilitates orientation analysis that indicates substantial texture weakening.

Figure 7.8 provides a detailed understanding of the twinning behavior and grain rotation in magnesium, a representative HCP metal, under compressive stress through shuffle and shear mechanisms. The figure showcases the transient HCP↔BCC transformation near the grain boundary region purely due to atomic shuffling under

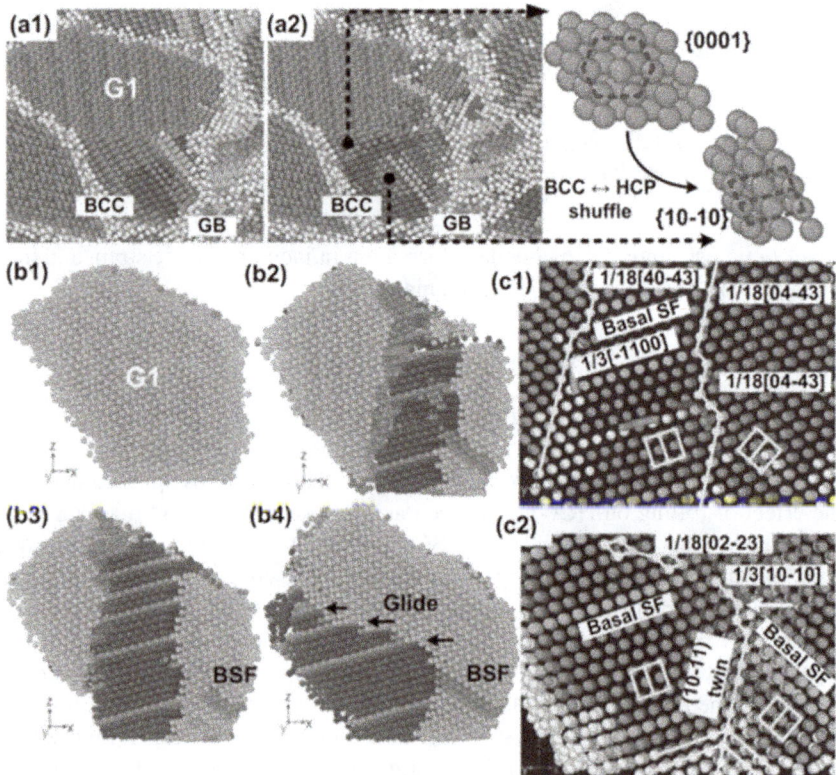

FIGURE 7.8 Snapshots of various stages of the deformation process in magnesium under compressive stress during the rolling process. The first set of snapshots (a1–a2) exhibits the BCC↔HCP shuffling transformation through polyhedral template matching (PTM). The second set (b1–b2) shows the formation of a twin region in the grain, and the third set (b3–b4) illustrates the dynamic traversing of the twin region along the grain under shear stress. The final set (c1–c2) displays the formation of interfacial disconnection steps through twinning dislocations (TDs) using PTM. The inset in the first set of snapshots highlights the initial {0001}HCP to final {10–10}HCP transformation.

compressive stress. As the specimen rolls along the rolling direction, the BCC phase grows into the grain interior but quickly transforms back into the HCP phase due to its metastable state, causing nucleation of the (10–11) compressive twin and altering the G1 grain's orientation. The orientation snapshots of the un-twinned and twinned regions of the G1 grain are presented in Figures 7.8(b1–b2). Furthermore, the formation of stacking faults in the twinned HCP region is also observed. The rolling process generates shear stress, causing the twin region to dynamically traverse along the G1 grain through interfacial disconnection steps facilitated by glissile zonal twinning dislocations. These results corroborate with previous experimental and simulation studies, validating the developed nano-rolling model.

Figure 7.9 presents a detailed analysis of the development of twin boundaries and boundary sliding faults during the rolling process, as well as the evolution of grain orientation and atomic stress. The investigation reveals that increasing the number of roller passes leads to a rise in the atomic fraction of both BSFs and TBs. The fraction of BSFs shows a significant increase after each roller pass, while the increase in TBs is relatively small after the second pass. This observation may be attributed to the presence of nano-sized grains in the specimen. However, the results also demonstrate that an increase in grain size causes a gradual enhancement of twin fraction, consistent with experimental findings. The effect of initial

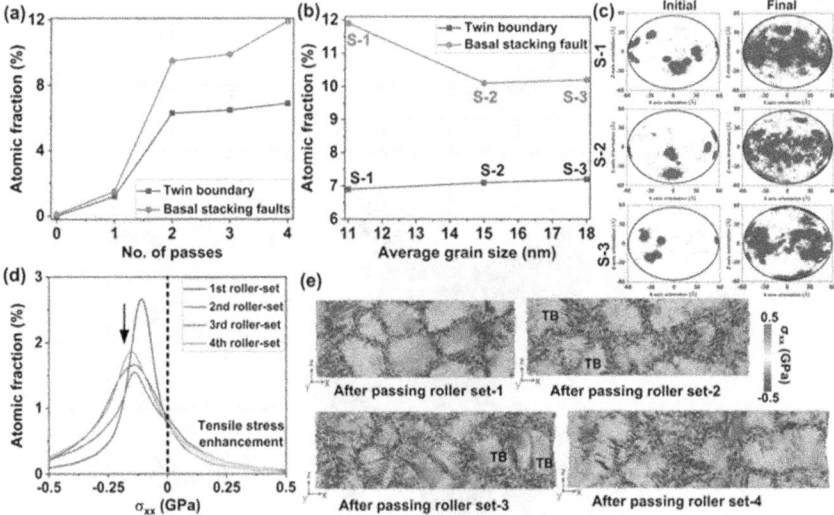

FIGURE 7.9 Results of the nano-rolling simulation study. (a and b) Plots showing the variation in the atomic fraction of twin boundaries (TBs) and basal stacking faults (BSFs) with respect to the number of rolling passes and the grain size, respectively. (c1–c6) Orientation snapshots of the three specimens with different grain sizes at initial and the final rolling conditions. (d) Plot of the atomic stress distribution along the rolling direction after the specimen traverses each roller set, where the shaded region indicates the tensile/positive stress in the specimen. Finally, (e) atomic stress snapshots show the evolution of local stresses in the specimen during the rolling process.

grain orientation on the final texture is also investigated, and the results indicate that all three specimens consistently show RD splitting after passing through the fourth set of rollers. The structural evolution of the specimen is greatly dependent on the atomic stress after each roller pass. The analysis reveals a decrease in peak stress values after the first roller pass and a small increment in positive stress in the specimen. However, increasing the number of passes marginally enhances the accumulation of compressive stress. The localized positive atomic stress is attributed to the twinning phenomenon in the specimen, as observed from the atomic stress snapshots, which is also substantiated in the existing literature. Overall, the results presented in this investigation provide valuable insights into the role of rolling on the development of TBs and BSFs, as well as the evolution of grain orientation and atomic stress. These findings are consistent with previous experimental and simulation studies and further reinforce the validity of the developed nano-rolling model. In summary, although Figure 7.9 demonstrates that increasing the number of roller passes increases the atomic fraction of BSFs and TBs, the atomic fraction of TBs remains almost constant due to the presence of nano-sized grains. However, a nominal increase in grain size enhances the twin fraction in the specimen, and the initial grain orientation also influences the final texture. All three specimens consistently show RD splitting after passing through the fourth set of rollers, which is dependent on the atomic stress.

MD simulation aids in the frame-by-frame exploration of the entire deformation process, and in this case, it especially gives insights into the structural evolution and corresponding orientation changes. In Figure 7.10, the authors analyze the orientation evolution of a single grain in a nano-crystalline Mg specimen during the texture weakening process. Quaternions are used to examine the atomic rotations of the grain. The results indicate that the basal texture weakens slightly due to the

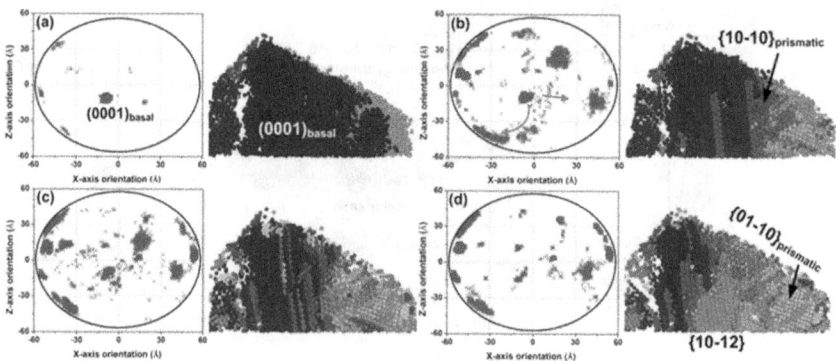

FIGURE 7.10 Structural evolution of a single grain in the NC Mg specimen through orientation plots and atomic snapshots. (a) Basal orientation of the grain, along with an atomic snapshot as inset. The results show that changing the basal orientation into non-basal planes causes texture weakening, as depicted in (b), where red arrows indicate the shift in orientation from {0001} basal to non-basal planes. Further, in (c), the initiation of {01–10} plane texture along with {10–10} planes is observed. Finally, (d) shows the evolution of stable {10–12} and {01–10} planes in the grain, which cause grain refinement and texture weakening.

formation of multiple layers of {10–10} prismatic slip planes that accommodate plastic deformation. This weakening is evident from the splitting of the basal orientation pole towards the ends of the x-axis. The study also shows the evolution of the stable {10–12} orientation, which transforms a single grain with basal orientation into multiple grains with different non-basal orientations. This transformation signifies simultaneous texture weakening and grain refinement. RD splitting suggests that texture weakening persists even after annealing. The study is consistent with previous literature on texture weakening and RD splitting in Mg. The authors also observe that grains with the pyramidal plane at the surface and normal to RD retain their initial grain orientation even after passing through the second set of rollers. BSFs are formed in the grain to accommodate the dynamic plastic strain generated during the compression process. Once the specimen passes through the rollers, most of these BSFs traverse towards the GBs with minimal structural variations. Figure 8.8 shows the formation of BSFs in a grain. Overall, the study provides valuable insights into the development of twinning and BSFs in Mg during nano-rolling deformation. The atomic fraction of BSFs and TBs increases with the number of roller passes, while the atomic fraction of TBs remains almost constant due to the presence of nano-sized grains. Grain size influences the twin fraction in the specimen. The initial grain orientation also influences the final texture, and RD splitting signifies simultaneous texture weakening and grain refinement.

7.2.4 Nano-Rolling Process of Metallic Glasses

The study of nano-rolling can be extended to metallic glasses to study the deformation behavior of these amorphous materials. Rolling processes are widely used to enhance the plasticity of metallic glasses by inducing a network of shear bands (SBs). However, the atomic-level understanding of the SB formation and propagation mechanism during mechanical processing is still limited. To address this issue, we developed an MD simulation model to recreate the rolling deformation process and investigate SB formation in Cu-Zr metallic glass specimens taken as representatives. The model was used to study the effect of temperature on the structural evolution, SB formation, atomic strain distribution, and atomic cluster formation. The simulation results were compared with previous experimental studies, and the findings were in good agreement. The model successfully predicted the rolling process of Cu-Zr metallic glass and precisely demonstrated the shear banding phenomenon. The study provides valuable insights into the mechanisms governing the SB formation and propagation during the rolling process of metallic glasses. The results highlight the importance of temperature in controlling structural evolution and SB formation in metallic glasses. Moreover, the atomic-level information obtained from the simulation can guide the development of more effective processing techniques to enhance the plasticity of metallic glasses. Overall, this section demonstrates the potential of MD simulations to investigate the structural evolution and deformation behavior of metallic glasses at the atomic level.

In this section, we provide a detailed analysis of the structural evolution in the MG specimen during different nano-rolling processes, such as cryo-, cold-, and hot-rolling. Figure 7.11 shows the atomic shear strain snapshots and the selective region strain distribution along the ND of the MG specimen during the cryo-rolling process.

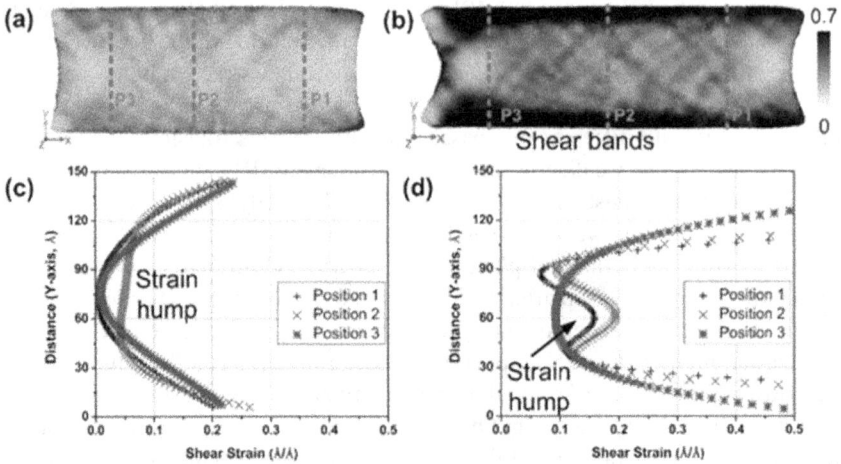

FIGURE 7.11 Atomic strain distribution of the Mg specimen cryo-rolled through the first and second sets of rollers. Atomic strain snapshots of the specimen are shown in panels (a) and (b), while panels (c) and (d) illustrate the atomic strain distribution plot along the ND at three representative positions after the first and second roller passes, respectively. The color scale in panels (a) and (b) represents the atomic shear strain value in angstroms per angstrom, highlighting the change in atomic position due to deformation.

The observations indicate that during the initial set of rollers, the density and distribution of the shear bands in the deformed specimen are low due to a smaller thickness reduction percentage. The SBs that do form are primarily oriented at an angle of approximately ~45°. However, as the deformed specimen is passed through the second set of rollers, the intensity of the SBs significantly increases, with strain accumulation occurring through the growth of previously formed bands. These findings are consistent with experimental analysis of mechanically processed metallic glass specimens through SEM. The strain distribution in the rolled specimen is shown at three representative regions: P1, P2, and P3, respectively. The strain is uniform, with increased strain at the top and bottom ends that gradually decreases near the center of the specimen at positions P1 and P3. At position P2, a non-uniform distribution of strain with a hump at the center of the specimen is present, indicating that the formation and propagation of SBs occur throughout the specimen. The curving in the front and back portion of the specimen suggests that the atoms near the surface (closer to the rollers) have traversed to a greater extent than the atoms at the central region. With an increase in the reduction percentage, higher strain is generated and stored in the previously formed and new SBs, resulting in a strain hump at multiple positions. The overall atomic shear strain in the MG specimen shows an increase in the strain shift towards the right side.

External forces applied to MG specimens can cause changes in their structural properties, as noted in a study by Luu et al. [2010]. To investigate these changes, we utilized the Voronoi tessellation method developed by Reddy and Pal [2017]. Our analysis involved tracking the alterations in the fraction of atomic clusters during the nano-rolling process. Figure 7.12 displays the plot of the change in various Voronoi

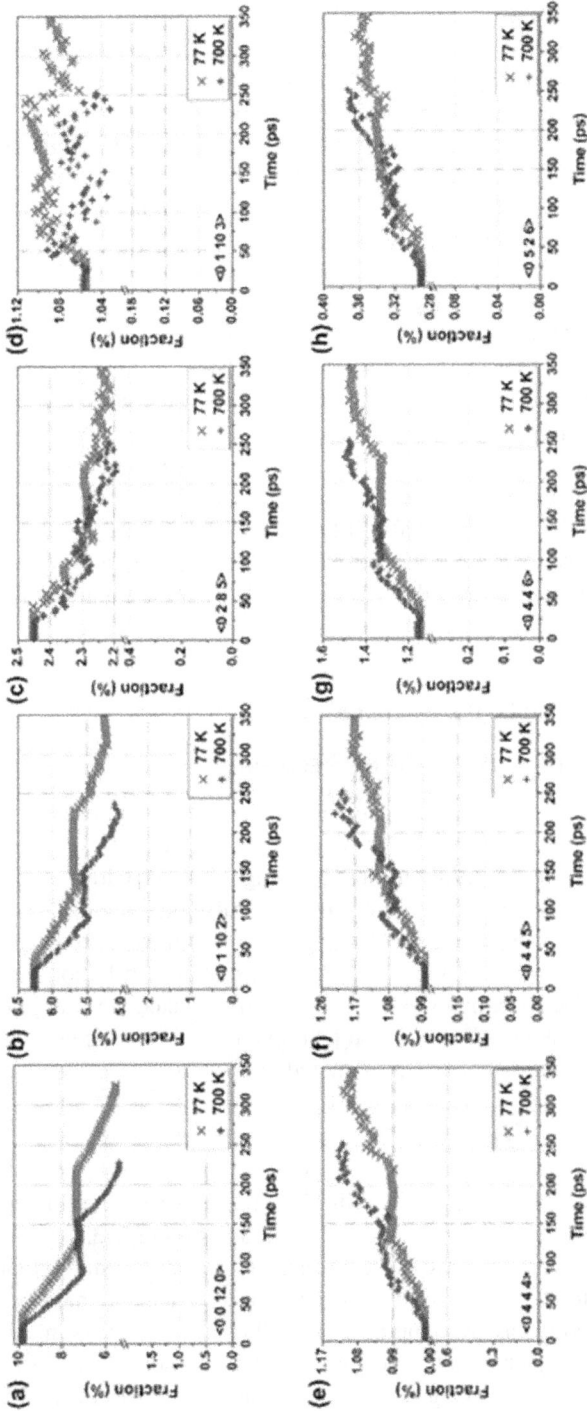

FIGURE 7.12 The plot shows the variation of various Voronoi clusters as a function of rolling time during cryo- and hot-rolling. The number fraction of different clusters is displayed for clusters (a) to (d) with high five-fold symmetry and for clusters (e) to (h) with a crystalline-like structure. The plot illustrates that the rate of decrease in ICO and ICO-like clusters and the rate of increase in crystalline-like clusters are slightly higher in hot-rolling compared to cryo-rolling.

clusters across the entire specimen as a function of rolling time for both cryo- and hot-rolling conditions. It's worth noting that the plateau region observed between the initial and final positions represents the time when the specimen has passed through the first set of rollers and is moving towards the second set, without being subjected to compressive force. During this period, the alteration in the fraction of Voronoi clusters remains constant. The results show that the icosahedral (ICO) and ICO-like clusters with high five-fold symmetry, such as <0 0 12 0>, <0 1 10 2>, and <0 1 10 3>, experience a declining tendency (with a constant value while moving from the first set to the second set of rollers), as depicted in Figures 7.12(a–d). Conversely, crystalline-like clusters like <0444>, <0445>, and <0446> display an increasing trend throughout the rolling process under both temperature conditions, as shown in Figures 7.12(e–h). Moreover, the decrease in the fraction of ICO and ICO-like clusters and the increase in crystalline-like clusters occur at a slightly higher rate in the case of hot-rolling as compared to cryo-rolling. The results of our investigation indicate that as the reduction percentage increases during the nano-rolling process, the metallic glass specimen may lose its glassy nature and begin to crystallize. This tendency is more pronounced at higher temperatures, as evidenced by the faster rate of decrease in icosahedral and icosahedral-like clusters and the faster rate of increase in crystalline-like clusters observed in hot-rolling compared to cryo-rolling. These findings suggest that stress-induced processes, such as rolling, can facilitate the transformation of metallic glass structures towards crystallization.

7.3   DESIGN OF ROLLING PROCESS USING LAMMPS

7.3.1   PREPARATION OF POLYCRYSTALLINE SPECIMEN

To begin, a single crystal metallic specimen with specific dimensions (in nm) was created using Atomsk [Hirel, 2015]. To create the nanocrystalline specimen, we employed the Voronoi tessellation method using Atomsk [Hirel, 2015], which involved dividing the single crystal specimen into several domains to form a representative volume element (RVE) made up of polygons. To facilitate domain formation, we generated a set of nucleating nodes with random spatial distribution throughout the simulation box. The specimen was then structurally relaxed via energy minimization using the conjugate gradient method [Plimpton, 1995] (as shown in step 2 of Figure 8.1). These nucleating nodes were denoted by the set:

$$\mathbf{g} = \left\{ g_1, g_2, \ldots\ldots\ldots g_m \right\}, g_i \in \mathbb{R}^n \tag{7.3}$$

Equation (7.3) indicates that all atoms within the space, $x \in \mathbb{R}n$, were linked to their nearest nucleating node, as illustrated in steps 2 and 3 of Figure 7.13.

After selecting the representative volume element, a seed crystallographic orientation is assigned to each polygon, and the orientation of each domain/polygon is modified while removing any overlapping atoms. This process results in the formation of grain boundaries, which ultimately leads to the creation of a nanostructured magnesium specimen with varying textures and grain sizes. Specifically, multiple grain sizes can be considered to produce statistically uniform results with a small

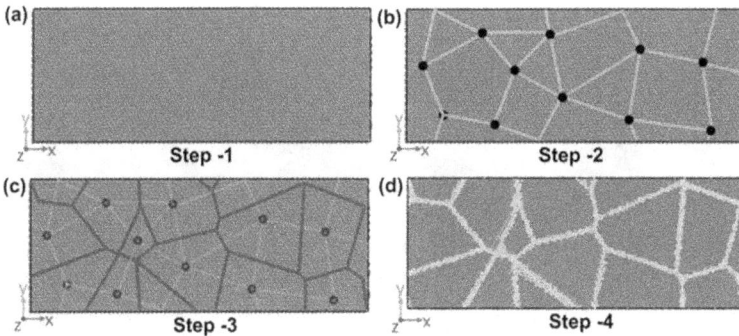

FIGURE 7.13 Step-by-step process of preparation of the nanocrystalline Mg specimen using the Voronoi tessellation method.

degree of error. To investigate the influence of initial orientation and grain size on the structural and texture evolution, the model is equilibrated at a certain temperature (user defined) under the NPT ensemble (where N represents the number of atoms, P represents pressure, and T represents temperature). The temperature is controlled using the Nosé–Hoover thermostat [Evans and Holian, 1985], and the equilibration process is simulated using LAMMPS [Plimpton, 1995]. To describe the interatomic interactions, an embedded atom method potential [Wilson and Mendelev, 2016] is utilized. These potentials can accurately predict bulk material properties such as the lattice constant, phase transitions, and melting temperatures and have previously been implemented for various deformation studies.

7.3.2 Model Setup of Roller-Specimen Assembly

The nano-rolling process involves merging the nanocrystalline specimen with sets of rollers (here, a representative four set of rollers is shown), forming the specimen-roller setup. Figure 7.14(a) illustrates an atomic snapshot of the NC metallic specimen along with the four sets of rollers. Meanwhile, Figures 7.14(b–d) show the grain size distribution of the three different NC metallic specimens, with average grain sizes of around 11, 15, and 18 nm, respectively. The user can increase or reduce the number of rollers depending on the thickness reduction percentage, material type, and operating temperature. The four sets of all-atom rigid cylindrical rollers, which have a close-packed crystal structure, are constructed with a diameter of approximately 14 nm and a total length of 15 nm (representative). The first set of rollers is positioned to produce a 10% thickness reduction, while the other three sets of rollers produce a 15% thickness reduction. To determine the reduction percentage for each roller set, the specimen that has passed through the previous set is taken into account. During roller rotation, the all-atom rollers interact with the specimen's atoms, causing shear stress. The roller sets rotate in opposite directions, with the upper rollers rotating anti-clockwise and the lower rollers rotating clockwise. The rolling direction (RD), normal direction (ND), and transverse direction (TD) are designated along the

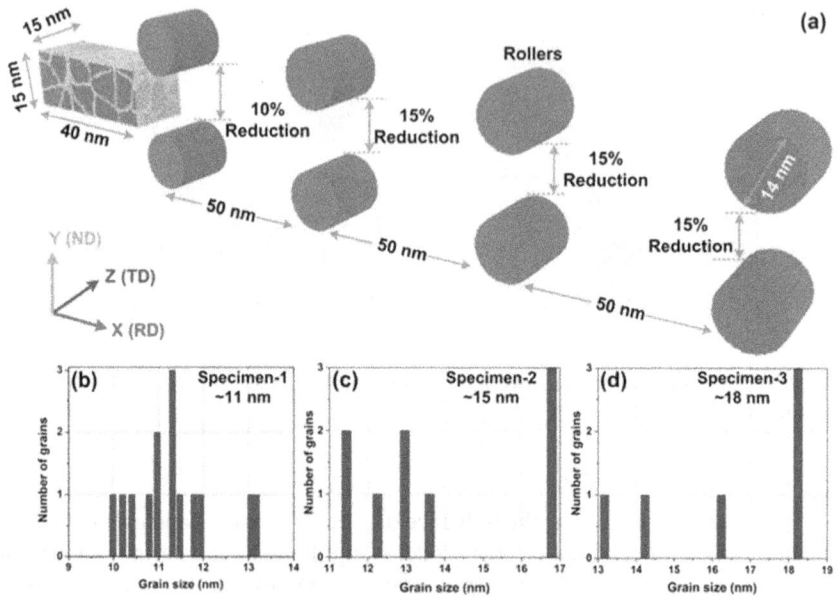

FIGURE 7.14 (a) Polyhedral template matching snapshots with dimensions of the NC Mg specimen along with the four set of the rollers; (b) plot showing the grain size distribution of the NC Mg (for three specimens with different initial grain sizes) used in the nano-rolling simulation.

x-, y-, and z-axes, respectively. The RD has non-periodic and shrink-wrapped boundary conditions, while the TD has periodic boundary conditions. The nano-rolling procedure is carried out under the NVT ensemble (where N is the number of particles, V is the volume, and T is the temperature). The OVITO post-processing package is used to visualize the structural deformation and orientation evolution of the nanocrystalline Mg specimen.

7.4 EXAMPLE LAMMPS INPUT CODE

 # This program is for obtaining rolling setup

```
units          metal
echo           both
atom_style     atomic
dimension      3
boundary       m m p
read_data      NC_Mg_equilibrated
lattice        hcp 3.20
region         slab block -199.2370089559 199.2370089559 -74.7138783585
        74.7138783585 -74.7138783585 74.7138783585 units box
group          slab2 region slab
lattice        hcp 3.20
```

```
region            cyl1 cylinder z 295 -145 70 -74.7138783585 74.7138783585
   units box
group             cyl1 region cyl1
create_atoms      1 region cyl1
lattice           hcp 3.20
region            cyl2 cylinder z 295 145 70 -74.7138783585 74.7138783585
   units box
group             cyl2 region cyl2
create_atoms      1 region cyl2
lattice           hcp 3.20
region            cyl3 cylinder z 795 -140 70 -74.7138783585 74.7138783585
   units box
group             cyl3 region cyl3
create_atoms      1 region cyl3
lattice           hcp 3.20
region            cyl4 cylinder z 795 140 70 -74.7138783585 74.7138783585
   units box
group             cyl4 region cyl4
create_atoms      1 region cyl4
region            slab1 block -203.2370089559 0 -77.7138783585 77.7138783585
   -74.7138783585 74.7138783585 units box
group             slab1 region slab1
region            slab3 block 0 203.2370089559 -77.7138783585 77.7138783585
   -74.7138783585 74.7138783585 units box
group             slab3 region slab3
group             cyl1 region cyl1
group             cyl2 region cyl2
group             cyl3 region cyl3
group             cyl4 region cyl4
group             boundary union slab1 slab3
group             rolls union cyl1 cyl2
group             rolls2 union cyl3 cyl4
#for region define
timestep          0.002
pair_style        eam/fs
pair_coeff        * * Mg1.eam.fs Mg
# Energy Minimization energy force
minimize          1.0e-8 1.0e-6 1000 1000
fix               equi boundary nve temp 300 300 0.01
run               1000
unfix             equi
compute              1 boundary stress/atom NULL
compute              2 boundary reduce sum c_1[1] c_1[2] c_1[3]
variable          stress equal ((c_2[3])/(3*vol))
variable          tmp equal xy
variable          lo equal ${tmp}
#variable         strain equal (xy-v_lo)/ly
velocity          all create 300 482748 rot yes mom yes dist gaussian
fix               54 rolls rigid single
fix               55 rolls2 rigid single
#fix              65 cyl2 rigid single
```

```
fix                 52 cyl1 move rotate 295.0 -145 73.322 0.0 0.0 -100.0 100 units box
fix                 64 cyl2 move rotate 295.0 145 73.322 0.0 0.0 100.0 100 units box
fix                 69 cyl3 move rotate 795.0 -140 73.322 0.0 0.0 -100.0 100 units box
fix                 71 cyl4 move rotate 795.0 140 73.322 0.0 0.0 100.0 100 units box
###----------------------------translation----------------------------###
fix                 94 boundary indent 100 cylinder z 295 -145 76 units box
fix                 95 boundary indent 100 cylinder z 295 145 76 units box
fix                 96 boundary indent 100 cylinder z 795 -140 76 units box
fix                 97 boundary indent 100 cylinder z 795 140 76 units box
fix                 4 boundary nvt temp 300 300 0.01
compute             KE all ke/atom
compute             peratom all pe/atom
thermo        200
thermo_style        custom step temp vol press pe ke enthalpy etotal
dump                1 all custom 1000 Mg_roll_dump_new*.lammpstrj id type x y z
dump                CSP all cfg 5000 dump.Mg_roll_dump@900K*.cfg mass type
    xs ys zs c_KE c_peratom fx fy fz
dump                33 all custom 5000 stress_peratom* id type x y z c_1[1] c_1[2]
    c_1[3] c_1[4] c_1[5] c_1[6]
log                 log_Mg_roll_new.data
fix                 push boundary addforce 0.003 0 0
run           400000
```

References

Abraham, F. F., Walkup, R., Gao, H., Duchaineau, M., Diaz De La Rubia, T., & Seager, M. (2002). Simulating materials failure by using up to one billion atoms and the world's fastest computer: Brittle fracture. *Proceedings of the National Academy of Sciences*, 99(9), 5777–5782.

Ahrens, J., Geveci, B., & Law, C. (2005). Paraview: An end-user tool for large data visualization. *The Visualization Handbook*, 717(8).

Aikin, R. M. (1997). The mechanical properties of in-situ composites. *JOM*, 49, 35–39.

Alder, B. J., & Wainwright, T. E. (1957). Phase transition for a hard sphere system. *The Journal of Chemical Physics*, 27(5), 1208–1209.

Amigo, N., Palominos, S., & Valencia, F. J. (2023). Machine learning modeling for the prediction of plastic properties in metallic glasses. *Scientific Reports*, 13(1), 348.

Antonov, S., Detrois, M., Isheim, D., Seidman, D., Helmink, R. C., Goetz, R. L., Sun, E., & Tin, S. (2015). Comparison of thermodynamic database models and APT data for strength modeling in high Nb content γ–γ′ Ni-base superalloys. *Materials & Design*, 86, 649–655.

Ashkenazy, Y., & Averback, R. S. (2012). Irradiation induced grain boundary flow—a new creep mechanism at the nanoscale. *Nano Letters*, 12(8), 4084–4089.

Basak, C. B., Sengupta, A. K., & Kamath, H. S. (2003). Classical molecular dynamics simulation of UO2 to predict thermophysical properties. *Journal of Alloys and Compounds*, 360(1–2), 210–216.

Bauri, R., & Yadav, D. (2018). Introduction to metal matrix composites. In *Metal Matrix Composites by Friction Stir Processing* (pp. 1–16), Oxford: Butterworth-Heinemann.

Berry, J., Rottler, J., Sinclair, C. W., & Provatas, N. (2015). Atomistic study of diffusion-mediated plasticity and creep using phase field crystal methods. *Physical Review B*, 92(13), 134103.

Bhattacharyya, D., Mara, N. A., Dickerson, P., Hoagland, R. G., & Misra, A. (2011). Compressive flow behavior of Al–TiN multilayers at nanometer scale layer thickness. *Acta Materialia*, 59(10), 3804–3816.

Borovikov, V., Mendelev, M. I., King, A. H., & LeSar, R. (2015). Effect of stacking fault energy on mechanism of plastic deformation in nanotwinned FCC metals. *Modelling and Simulation in Materials Science and Engineering*, 23(5), 055003.

Casati, R., & Vedani, M. (2014). Metal matrix composites reinforced by nano-particles—a review. *Metals (Basel)*, 4, 65–83.

Chavoshi, S. Z., Xu, S., & Goel, S. (2017). Addressing the discrepancy of finding the equilibrium melting point of silicon using molecular dynamics simulations. *Proceedings of the Royal Society A: Mathematical, Physical and Engineering Sciences*, 473(2202), 20170084.

Chawla, K. K. (2012). *Composite Materials*. New York, NY: Springer.

Chen, D. (1995). Structural modeling of nanocrystalline materials. *Computational Materials Science*, 3(3), 327–333.

Chen, M., Ma, E., Hemker, K. J., Sheng, H., Wang, Y., & Cheng, X. (2003). Deformation twinning in nanocrystalline aluminium. *Science*, 300(5623), 1275–1277.

Choi, B. K., Yoon, G. H., & Lee, S. (2016). Molecular dynamics studies of CNT-reinforced aluminum composites under uniaxial tensile loading. *Composites Part B: Engineering*, 91, 119–125.

Choi, H. J., Min, B. H., Shin, J. H., & Bae, D. H. (2011). Strengthening in nanostructured 2024 aluminum alloy and its composites containing carbon nanotubes. *Composites Part A: Applied Science and Manufacturing*, 42(10), 1438–1444.

Chunfeng, D., Zhang, X., Yanxia, M. A., & Dezun, W. (2007). Fabrication of aluminum matrix composite reinforced with carbon nanotubes. *Rare Metals*, 26(5), 450–455.

Coleman, S. P., Sichani, M. M., & Spearot, D. E. (2014). A computational algorithm to produce virtual X-ray and electron diffraction patterns from atomistic simulations. *Jom*, 66, 408–416.

Coleman, S. P., Spearot, D. E., & Capolungo, L. (2013). Virtual diffraction analysis of Ni [0 1 0] symmetric tilt grain boundaries. *Modelling and Simulation in Materials Science and Engineering*, 21(5), 055020.

Dandekar, C. R., & Shin, Y. C. (2011). Molecular dynamics based cohesive zone law for describing Al-SiC interface mechanics. *Composites Part A: Applied Science and Manufacturing*, 42(4), 355–363.

Deng, B., Hsu, P. C., Chen, G., Chandrashekar, B. N., Liao, L., Ayitimuda, Z., Wu, J., Guo, Y., Lin, L., Zhou, Y., Aisijiang, M., Xie, Q., Cui, Y., Liu, Z., & Peng, H. (2015). Roll-to-roll encapsulation of metal nanowires between graphene and plastic substrate for high-performance flexible transparent electrodes. *Nano Letters*, 15(6), 4206–4213.

Derlet, P. M., Hasnaoui, A., & Van Swygenhoven, H. (2003). Atomistic simulations as guidance to experiments. *Scripta Materialia*, 49(7), 629–635.

Desai, T. G., Millett, P., & Wolf, D. (2008). Is diffusion creep the cause for the inverse Hall–Petch effect in nanocrystalline materials? *Materials Science and Engineering: A*, 493(1–2), 41–47.

Dieter, G. E., & Bacon, D. J. (1986). *Mechanical Metallurgy*, vol. 3. New York: McGraw-Hill.

Dillon, S. J., Tang, M., Carter, W. C., & Harmer, M. P. (2007). Complexion: A new concept for kinetic engineering in materials science. *Acta Materialia*, 55(18), 6208–6218.

Evans, D. J., & Holian, B. L. (1985). The nose–hoover thermostat. *The Journal of Chemical Physics*, 83(8), 4069–4074.

Faken, D., & Jónsson, H. (1994). Systematic analysis of local atomic structure combined with 3D computer graphics. *Computational Materials Science*, 2(2), 279–286.

Falk, M. L., & Langer, J. S. (1998). Dynamics of viscoplastic deformation in amorphous solids. *Physical Review E*, 57(6), 7192.

Foroughi, A., Tavakoli, R., & Aashuri, H. (2016). Molecular dynamics study of structural formation in Cu50–Zr50 bulk metallic glass. *Journal of Non-Crystalline Solids*, 432, 334–341.

Frøseth, A. G., Derlet, P. M., & Van Swygenhoven, H. (2004). Dislocations emitted from nanocrystalline grain boundaries: Nucleation and splitting distance. *Acta Materialia*, 52(20), 5863–5870.

Ghosh, S., & Chokshi, A. H. (2014). Creep in nanocrystalline zirconia. *Scripta Materialia*, 86, 13–16.

Goswami, D., Munera, J. C., Pal, A., Sadri, B., Scarpetti, C. L. P., & Martinez, R. V. (2018). Roll-to-roll nanoforming of metals using laser-induced superplasticity. *Nano Letters*, 18(6), 3616–3622.

Gotman, I., Koczak, M. J., & Shtessel, E. (1994). Fabrication of Al matrix *in situ* composites via self-propagating synthesis. *Materials Science and Engineering: A*, 187, 189–199.

Gupta, V. K., Yoon, D. H., Meyer III, H. M., & Luo, J. (2007). Thin intergranular films and solid-state activated sintering in nickel-doped tungsten. *Acta Materialia*, 55(9), 3131–3142.

Haghshenas, M. (2016). Metal-matrix composites. In *Reference Module in Materials Science and Materials Engineering*. Oxford: Elsevier.

Han, S. M., Phillips, M. A., & Nix, W. D. (2009). Study of strain softening behavior of Al–Al3Sc multilayers using microcompression testing. *Acta Materialia*, 57(15), 4473–4490.

Herron, A. D., Coleman, S. P., Dang, K. Q., Spearot, D. E., & Homer, E. R. (2018). Simulation of kinematic Kikuchi diffraction patterns from atomistic structures. *MethodsX*, 5, 1187–1203.

Hirata, A., Kang, L. J., Fujita, T., Klumov, B., Matsue, K., Kotani, M., Yavari, A. R., & Chen, M. W. (2013). Geometric frustration of icosahedron in metallic glasses. *Science*, 341(6144), 376–379.

Hirel, P. (2015). Atomsk: A tool for manipulating and converting atomic data files. *Computer Physics Communications*, 197, 212–219.

Hirth, J. P., & Lothe, J. (1992). *Theory of Dislocations*, 2nd ed. Malabar, UK: Krieger Publishing.

Hondros, E. D., & Seah, M. P. (1977). Segregation to interfaces. *International Metals Reviews*, 22(1), 262–301.

Honeycutt, J. D., & Andersen, H. C. (1987). Molecular dynamics study of melting and freezing of small Lennard-Jones clusters. *Journal of Physical Chemistry*, 91(19), 4950–4963.

Hull, D., & Clyne, T. W. (1996). *An Introduction to Composite Materials*. Cambridge: Cambridge University Press.

Humphrey, W., Dalke, A., & Schulten, K. (1996). VMD: Visual molecular dynamics. *Journal of Molecular Graphics*, 14(1), 33–38.

Jacoby, M. (2004). *Composite Materials*. Chicago, IL: Elsevier.

Jafari, M., Abbasi, M. H., Enayati, M. H., & Karimzadeh, F. (2012). Mechanical properties of nanostructured Al2024-MWCNT composite prepared by optimized mechanical milling and hot pressing methods. *Advanced Powder Technology*, 23(2), 205–210.

Ji, C., Shi, J., Liu, Z., & Wang, Y. (2013). Comparison of tool–chip stress distributions in nano-machining of monocrystalline silicon and copper. *International Journal of Mechanical Sciences*, 77, 30–39.

Jud, E., Huwiler, C. B., & Gauckler, L. J. (2005). Sintering analysis of undoped and cobalt oxide doped ceria solid solutions. *Journal of the American Ceramic Society*, 88(11), 3013–3019.

Judelewicz, M., Künzi, H. U., Merk, N., & Ilschner, B. (1994). Microstructural development during fatigue of copper foils 20–100 μm thick. *Materials Science and Engineering: A*, 186(1–2), 135–142.

Kadau, K., Germann, T. C., & Lomdahl, P. S. (2006). Molecular dynamics comes of age: 320 billion atom simulation on BlueGene/L. *International Journal of Modern Physics C*, 17(12), 1755–1761.

Karkina, L., Karkin, I., Kuznetsov, A., & Gornostyrev, Y. (2019). Alloying element segregation and grain boundary reconstruction, atomistic modeling. *Metals*, 9(12), 1319.

Keblinski, P., Wolf, D., & Gleiter, H. (1998). Molecular-dynamics simulation of grain-boundary diffusion creep. *Interface Science*, 6, 205–212.

Kelchner, C. L., Plimpton, S. J., & Hamilton, J. C. (1998). Dislocation nucleation and defect structure during surface indentation. *Physical Review B*, 58(17), 11085.

Klement, W., Willens, R. H., & Duwez, P. O. L. (1960). Non-crystalline structure in solidified gold–silicon alloys. *Nature*, 187(4740), 869–870.

Kobayashi, S., Kamata, A., & Watanabe, T. (2009). Roles of grain boundary microstructure in high-cycle fatigue of electrodeposited nanocrystalline Ni–P alloy. *Scripta Materialia*, 61(11), 1032–1035.

Koilraj, T. T., & Kalaichelvan, K. (2015). Hybrid nanocomposites—a review. *Applied Mechanics and Materials*, 766–767, 50–56.

Koleini, M. M., Badizad, M. H., Ghatee, M. H., & Ayatollahi, S. (2019). An atomistic insight into the implications of ion-tuned water injection in wetting preferences of carbonate reservoirs. *Journal of Molecular Liquids*, 293, 111530.

Krakow, R., Bennett, R. J., Johnstone, D. N., Vukmanovic, Z., Solano-Alvarez, W., Lainé, S. J., . . . & Hielscher, R. (2017). On three-dimensional misorientation spaces. *Proceedings of the Royal Society A: Mathematical, Physical and Engineering Sciences*, 473(2206), 20170274.

Kruzic, J. J. (2016). Bulk metallic glasses as structural materials: A review. *Advanced Engineering Materials*, 18(8), 1308–1331.

Kwon, H., Estili, M., Takagi, K., Miyazaki, T., & Kawasaki, A. (2009). Combination of hot extrusion and spark plasma sintering for producing carbon nanotube reinforced aluminum matrix composites. *Carbon*, 47(3), 570–577.

Kwon, H., Park, D. H., Silvain, J. F., & Kawasaki, A. (2010). Investigation of carbon nanotube reinforced aluminum matrix composite materials. *Composites Science and Technology*, 70(3), 546–550.

Larsen, P. M., Schmidt, S., & Schiøtz, J. (2016). Robust structural identification via polyhedral template matching. *Modelling and Simulation in Materials Science and Engineering*, 24(5), 055007.

Lee, J. G. (2016). *Computational Materials Science: An Introduction*. Boca Raton: CRC Press.

Lennard-Jones, J. E. (1925). On the forces between atoms and ions. *Proceedings of the Royal Society of London. Series A, Containing Papers of a Mathematical and Physical Character*, 109(752), 584–597.

Li, H., & Ebrahimi, F. (2003). Synthesis and characterization of electrodeposited nanocrystalline nickel-iron alloys. *Materials Science and Engineering: A*, 347(1–2), 93–101.

Li, J. (2003). AtomEye: An efficient atomistic configuration viewer. *Modelling and Simulation in Materials Science and Engineering*, 11(2), 173.

Liu, Z. Y., Xiao, B. L., Wang, W. G., & Ma, Z. Y. T (2014). Tensile strength and electrical conductivity of carbon nanotube reinforced aluminum matrix composites fabricated by powder metallurgy combined with friction stir processing. *Journal of Materials Science & Technology*, 30(7), 649–655.

Lodesani, F., Menziani, M. C., Hijiya, H., Takato, Y., Urata, S., & Pedone, A. (2020). Structural origins of the mixed alkali effect in alkali aluminosilicate glasses: Molecular dynamics study and its assessment. *Scientific Reports*, 10(1), 1–18.

Lu, Q., & Bhattacharya, B. (2005). Effect of randomly occurring Stone–Wales defects on mechanical properties of carbon nanotubes using atomistic simulation. *Nanotechnology*, 16(4), 555.

Lu, Y. B., Yang, Q. S., He, X. Q., & Liew, K. M. (2016). Modeling the interfacial behavior of carbon nanotube fiber/polyethylene composites by molecular dynamics approach. *Computational Materials Science*, 114, 189–198.

Luo, J., Cheng, H., Asl, K. M., Kiely, C. J., & Harmer, M. P. (2011). The role of a bilayer interfacial phase on liquid metal embrittlement. *Science*, 333(6050), 1730–1733.

Luu, H. T., Ravelo, R. J., Rudolph, M., Bringa, E. M., Germann, T. C., Rafaja, D., & Gunkelmann, N. (2020). Shock-induced plasticity in nanocrystalline iron: Large-scale molecular dynamics simulations. *Physical Review B*, 102(2), 020102.

Malaki, M., Xu, W., Kasar, A., Menezes, P., Dieringa, H., Varma, R., & Gupta, M. (2019). Advanced metal matrix nanocomposites. *Metals (Basel)*, 9, 330.

Mara, N. A., Bhattacharyya, D., Hirth, J. P., Dickerson, P., & Misra, A. (2010). Mechanism for shear banding in nanolayered composites. *Applied Physics Letters*, 97(2), 021909.

Marinica, M. C., Ventelon, L., Gilbert, M. R., Proville, L., Dudarev, S. L., Marian, J., . . . & Willaime, F. (2013). Interatomic potentials for modelling radiation defects and dislocations in tungsten. *Journal of Physics: Condensed Matter*, 25(39), 395502.

McCabe, R. J., Beyerlein, I. J., Carpenter, J. S., & Mara, N. A. (2014). The critical role of grain orientation and applied stress in nanoscale twinning. *Nature Communications*, 5(1), 1–7.

Mendelev, M. I., Kramer, M. J., Becker, C. A., & Asta, M. (2008). Analysis of semi-empirical interatomic potentials appropriate for simulation of crystalline and liquid Al and Cu. *Philosophical Magazine*, 88(12), 1723–1750.

Meraj, M., & Pal, S. (2016). Deformation of Ni20W20Cu20Fe20Mo20 high entropy alloy for tensile followed by compressive and compressive followed by tensile loading: A molecular dynamics simulation-based study. *IOP Conference Series: Materials Science and Engineering*, 115(1), 012019.

Meraj, M., & Pal, S. (2019). Molecular dynamics-based study of high temperature deformation process of nanocrystalline Ni-Nb alloy under tensile loading condition. *Materials Today: Proceedings*, 11, 740–746.

Meyers, M. A., Mishra, A., & Benson, D. J. (2006). Mechanical properties of nanocrystalline materials. *Progress in Materials Science*, 51(4), 427–556.

Mianroodi, J. R., Hunter, A., Beyerlein, I. J., & Svendsen, B. (2016). Theoretical and computational comparison of models for dislocation dissociation and stacking fault/core formation in fcc crystals. *Journal of the Mechanics and Physics of Solids*, 95, 719–741.

Mishra, S., & Pal, S. (2018). Variation of glass transition temperature of Al90Sm10 metallic glass under pressurized cooling. *Journal of Non-Crystalline Solids*, 500, 249–259.

Mitra, R., Hoffman, R. A., Madan, A., & Weertman, J. R. (2001). Effect of process variables on the structure, residual stress, and hardness of sputtered nanocrystalline nickel films. *Journal of Materials Research*, 16(4), 1010–1027.

Miyamoto, M., Nishijima, D., Ueda, Y., Doerner, R. P., Kurishita, H., Baldwin, M. J., Morito, S., Ono, K., & Hanna, J. (2009). Observations of suppressed retention and blistering for tungsten exposed to deuterium–helium mixture plasmas. *Nuclear Fusion*, 49(6), 065035.

Murashkin, M. Y., Sabirov, I., Medvedev, A. E., Enikeev, N. A., Lefebvre, W., Valiev, R. Z., & Sauvage, X. (2016). Mechanical and electrical properties of an ultrafine grained Al–8.5 wt.% RE (RE= 5.4 wt.% Ce, 3.1 wt.% La) alloy processed by severe plastic deformation. *Materials & Design*, 90, 433–442.

Nie, J., Chan, J. M., Qin, M., Zhou, N., & Luo, J. (2017). Liquid-like grain boundary complexion and sub-eutectic activated sintering in CuO-doped TiO2. *Acta Materialia*, 130, 329–338.

Odnobokova, M., Belyakov, A., & Kaibyshev, R. (2015). Development of nanocrystalline 304L stainless steel by large strain cold working. *Metals*, 5(2), 656–668.

Omanović-Mikličanin, E., Badnjević, A., Kazlagić, A., & Hajlovac, M. (2020). Nanocomposites: A brief review. *Health and Technology (Berlin)*, 10, 51–59.

Ostovan, F., Matori, K. A., Toozandehjani, M., Oskoueian, A., Yusoff, H. M., Yunus, R., Ariff, A. H. M., Quah, H. J., & Lim, W. F. (2015). Effects of CNTs content and milling time on mechanical behavior of MWCNT-reinforced aluminum nanocomposites. *Materials Chemistry and Physics*, 166, 160–166.

Pal, S., Meraj, M., & Deng, C. (2017). Effect of Zr addition on creep properties of ultra-fine grained nanocrystalline Ni studied by molecular dynamics simulations. *Computational Materials Science*, 126, 382–392.

Pal, S., Reddy, K. V., & Deng, C. (2019). On the role of Cu-Zr amorphous intergranular films on crack growth retardation in nanocrystalline Cu during monotonic and cyclic loading conditions. *Computational Materials Science*, 169, 109122.

Pal, S., Reddy, K. V., & Spearot, D. E. (2020). Zr segregation in Ni–Zr alloy: Implication on deformation mechanism during shear loading and bending creep. *Journal of Materials Science*, 55, 6172–6186.

Pal, S., Reddy, K. V., Yu, T., Xiao, J., & Deng, C. (2021). The spectrum of atomic excess free volume in grain boundaries. *Journal of Materials Science*, 56, 11511–11528.

Plimpton, S. (1995). Fast parallel algorithms for short-range molecular dynamics. *Journal of Computational Physics*, 117(1), 1–19.

Probst, P. T., Sekar, S., König, T. A., Formanek, P., Decher, G., Fery, A., & Pauly, M. (2018). Highly oriented nanowire thin films with anisotropic optical properties driven by the simultaneous influence of surface templating and shear forces. *ACS Applied Materials & Interfaces*, 10(3), 3046–3057.

Raabe, D., Herbig, M., Sandlöbes, S., Li, Y., Tytko, D., Kuzmina, M., Ponge, D., & Choi, P. P. (2014). Grain boundary segregation engineering in metallic alloys: A pathway to the design of interfaces. *Current Opinion in Solid State and Materials Science*, 18(4), 253–261.

Read, D. T. (1998). Tension-tension fatigue of copper thin films. *International Journal of Fatigue*, 20(3), 203–209.

Reddy, K. V., Deng, C., & Pal, S. (2019). Intensification of shock damage through heterogeneous phase transition and dislocation loop formation due to presence of pre-existing line defects in single crystal Cu. *Journal of Applied Physics*, 126(17), 174302.

Reddy, K. V., Meraj, M., & Pal, S. (2017). Mechanistic study of bending creep behaviour of bicrystal nanobeam. *Computational Materials Science*, 136, 36–43.

Reddy, K. V., & Pal, S. (2017). Contribution of Nb towards enhancement of glass forming ability and plasticity of Ni-Nb binary metallic glass. *Journal of Non-Crystalline Solids*, 471, 243–250.

Reddy, K. V., & Pal, S. (2018a). Effect of grain boundary complexions on the deformation behavior of Ni bicrystal during bending creep. *Journal of Molecular Modeling*, 24, 1–12.

Reddy, K. V., & Pal, S. (2018b). Influence of dislocations, twins, and stacking faults on the fracture behavior of nanocrystalline Ni nanowire under constant bending load: A molecular dynamics study. *Journal of Molecular Modeling*, 24, 1–11.

Reddy, K. V., & Pal, S. (2018c). Influence of grain boundary complexion on deformation mechanism of high temperature bending creep process of Cu bicrystal. *Transactions of the Indian Institute of Metals*, 71(7), 1721–1734.

Reddy, K. V., & Pal, S. (2019). Evaluation of glass forming ability of Zr–Nb alloy systems through liquid fragility and Voronoi cluster analysis. *Computational Materials Science*, 158, 324–332.

Reddy, K. V., & Pal, S. (2020). Shock velocity-dependent elastic-plastic collapse of pre-existing stacking fault tetrahedron in single crystal Cu. *Computational Materials Science*, 172, 109390.

Rohatgi, B. P. P. K., Schultz, B., & Matters, M. (2011). Lightweight metal matrix nanocomposites—stretching the boundaries of metals. *Material Matters*, 1–6.

Schäfer, J., & Albe, K. (2012). Influence of solutes on the competition between mesoscopic grain boundary sliding and coupled grain boundary motion. *Scripta Materialia*, 66(5), 315–317.

Scudino, S., & Şopu, D. (2018). Strain distribution across an individual shear band in real and simulated metallic glasses. *Nano Letters*, 18(2), 1221–1227.

Seah, M. P. (1980). Grain boundary segregation. *Journal of Physics F: Metal Physics*, 10(6), 1043.

Senkov, O. N. (2007). Correlation between fragility and glass-forming ability of metallic alloys. *Physical Review B*, 76(10), 104202.

Sheibani, S., & Najafabadi, M. F. (2007). In situ fabrication of Al-TiC metal matrix composites by reactive slag process. *Materials & Design*, 28, 2373–2378.

Sheng, H., Şopu, D., Fellner, S., Eckert, J., & Gammer, C. (2022). Mapping shear bands in metallic glasses: From atomic structure to bulk dynamics. *Physical Review Letters*, 128(24), 245501.

Shi, P., Engström, A., Höglund, L., Sundman, B., & Ågren, J. (2005). Thermo-calc and DICTRA enhance materials design and processing. *Materials Science Forum*, 475, 3339–3346.

Shimizu, F., Ogata, S., & Li, J. (2007). Theory of shear banding in metallic glasses and molecular dynamics calculations. *Materials Transactions*, 48(11), 2923–2927.

Soklaski, R., Nussinov, Z., Markow, Z., Kelton, K. F., & Yang, L. (2013). Connectivity of icosahedral network and a dramatically growing static length scale in Cu-Zr binary metallic glasses. *Physical Review B*, 87(18), 184203.

Srivastava, A. (2017). Metal matrix nanocomposites (MMCs): A review of their physical and mechanical properties. *International Journal of Nanotechnology in Medicine & Engineering*, 2, 152–154.

Stukowski, A. (2009). Visualization and analysis of atomistic simulation data with OVITO–the open visualization tool. *Modelling and Simulation in Materials Science and Engineering*, 18(1), 015012.

Stukowski, A., Bulatov, V. V., & Arsenlis, A. (2012). Automated identification and indexing of dislocations in crystal interfaces. *Modelling and Simulation in Materials Science and Engineering*, 20(8), 085007.

Subramaniam, A., Koch, C. T., Cannon, R. M., & Rühle, M. (2006). Intergranular glassy films: An overview. *Materials Science and Engineering: A*, 422(1–2), 3–18.

Suwas, S., & Ray, R. K. (2014). *Crystallographic Texture of Materials* (pp. 108–173). London: Springer.

Tang, C., & Wong, C. H. (2015). A molecular dynamics simulation study of solid-like and liquid-like networks in Zr46Cu46Al8 metallic glass. *Journal of Non-Crystalline Solids*, 422, 39–45.

Tang, M., Carter, W. C., & Cannon, R. M. (2006). Grain boundary order-disorder transitions. *Journal of Materials Science*, 41, 7691–7695.

Taskaev, S., Skokov, K., Khovaylo, V., Buchelnikov, V., Pellenen, A., Karpenkov, D., Ulyanov, M., Bataev, D., Usenko, A., Lyange, M., & Gutfleisch, O. (2015). Effect of severe plastic deformation on the specific heat and magnetic properties of cold rolled Gd sheets. *Journal of Applied Physics*, 117(12), 123914.

Tersoff, J. (1988). Empirical interatomic potential for carbon, with applications to amorphous carbon. *Physical Review Letters*, 61(25), 2879.

Thomas, S., & Lee, S. U. (2019). Atomistic insights into the anisotropic mechanical properties and role of ripples on the thermal expansion of h-BCN monolayers. *RSC Advances*, 9(3), 1238–1246.

Trady, S., Hasnaoui, A., & Mazroui, M. (2017). Atomic packing and medium-range order in Ni3Al metallic glass. *Journal of Non-Crystalline Solids*, 468, 27–33.

Trady, S., Mazroui, M., Hasnaoui, A., & Saadouni, K. (2016). Molecular dynamics study of atomic-level structure in monatomic metallic glass. *Journal of Non-Crystalline Solids*, 443, 136–142.

Tranh, D. T. N., Van Hoang, V., & Hanh, T. T. T. (2021). Modeling glassy SiC nanoribbon by rapidly cooling from the liquid: An affirmation of appropriate potentials. *Physica B: Condensed Matter*, 608, 412746.

Trautt, Z. T., Adland, A., Karma, A., & Mishin, Y. (2012). Coupled motion of asymmetrical tilt grain boundaries: Molecular dynamics and phase field crystal simulations. *Acta Materialia*, 60(19), 6528–6546.

Tsai, P. C., Jeng, Y. R., Lee, J. T., Stachiv, I., & Sittner, P. (2017). Effects of carbon nanotube reinforcement and grain size refinement mechanical properties and wear behaviors of carbon nanotube/copper composites. *Diamond and Related Materials*, 74, 197–204.

Turnbull, D. (1961). The 1961 Institute of metals division lecture-the liquid state and the liquid-solid transition. *Transactions of the Metallurgical Society of AIME*, 221(3), 422–439.

Varma, S. K., Caballero, V., Ponce, J., De La Cruz, A., & Salas, D. (1996). The effect of stacking fault energy on the microstructural development during room temperature wire drawing in Cu, Al and their dilute alloys. *Journal of Materials Science*, 31(21), 5623–5630.

Wang, C. H., Chao, K. C., Fang, T. H., Stachiv, I., & Hsieh, S. F. (2016). Investigations of the mechanical properties of nanoimprinted amorphous Ni–Zr alloys utilizing the molecular dynamics simulation. *Journal of Alloys and Compounds*, 659, 224–231.

Wang, D., Volkert, C. A., & Kraft, O. (2008). Effect of length scale on fatigue life and damage formation in thin Cu films. *Materials Science and Engineering: A*, 493(1–2), 267–273.

Watanabe, T., & Tsurekawa, S. (1999). The control of brittleness and development of desirable mechanical properties in polycrystalline systems by grain boundary engineering. *Acta Materialia*, 47(15–16), 4171–4185.

Wei-Zhong, L., Cong, C., & Jian, Y. (2008). Molecular dynamics simulation of self-diffusion coefficient and its relation with temperature using simple Lennard-Jones potential. *Heat Transfer—Asian Research*, 37(2), 86–93.

Welsch, E., Ponge, D., Haghighat, S. H., Sandlöbes, S., Choi, P., Herbig, M., Zaefferer, S., & Raabe, D. (2016). Strain hardening by dynamic slip band refinement in a high-Mn lightweight steel. *Acta Materialia*, 116, 188–199.

Wilde, G., & Divinski, S. (2019). Grain boundaries and diffusion phenomena in severely deformed materials. *Materials Transactions*, 60(7), 1302–1315.

Wilson, S. R., & Mendelev, M. I. (2016). A unified relation for the solid-liquid interface free energy of pure FCC, BCC, and HCP metals. *The Journal of Chemical Physics*, 144(14).

Wu, H. C., Kumar, A., Wang, J., Bi, X. F., Tomé, C. N., Zhang, Z., & Mao, S. X. (2016). Rolling-induced face centered cubic titanium in hexagonal close packed titanium at room temperature. *Scientific Reports*, 6(1), 1–8.

Wu, S. J., Liu, Z. Q., Qu, R. T., & Zhang, Z. F. (2021). Designing metallic glasses with optimal combinations of glass-forming ability and mechanical properties. *Journal of Materials Science & Technology*, 67, 254–264.

Xiang, J., Xie, L., Meguid, S. A., Pang, S., Yi, J., Zhang, Y., & Liang, R. (2017). An atomic-level understanding of the strengthening mechanism of aluminum matrix composites reinforced by aligned carbon nanotubes. *Computational Materials Science*, 128, 359–372.

Xu, F., Fang, F., Zhu, Y., & Zhang, X. (2017). Study on crystallographic orientation effect on surface generation of aluminum in nano-cutting. *Nanoscale Research Letters*, 12(1), 1–13.

Xu, W., & Davila, L. P. (2017). Size dependence of elastic mechanical properties of nanocrystalline aluminum. *Materials Science and Engineering: A*, 692, 90–94.

Yan, R., Ma, S. D., Sun, W. Z., Jing, T., & Dong, H. B. (2020). The solid–liquid interface free energy of Al: A comparison between molecular dynamics calculations and experimental measurements. *Computational Materials Science*, 184, 109910.

Yildiz, Y. O., & Kirca, M. (2018). Compression and shear behavior of ultrathin coated nanoporous gold: A molecular dynamics study. *Journal of Applied Physics*, 124(18).

Yoshino, M., Umehara, N., & Aravindan, S. (2009). Development of functional surface by nano-plastic forming. *Wear*, 266(5–6), 581–584.

Yuan, Y., Greuner, H., Böswirth, B., Linsmeier, C., Luo, G. N., Fu, B. Q., Xu, H. Y., Shen, Z. J., & Liu, W. (2013). Surface modification of molten W exposed to high heat flux helium neutral beams. *Journal of Nuclear Materials*, 437(1–3), 297–302.

Zhang, X., Bao, L., Wu, Y. Y., Zhu, X. L., & Tan, Z. J. (2017). Radial distribution function of semiflexible oligomers with stretching flexibility. *The Journal of Chemical Physics*, 147(5).

Zhong, C., Zhang, H., Cao, Q. P., Wang, X. D., Zhang, D. X., Ramamurty, U., & Jiang, J. Z. (2016). Deformation behavior of metallic glasses with shear band like atomic structure: A molecular dynamics study. *Scientific Reports*, 6(1), 30935.

Index

For Product Safety Concerns and Information please contact our EU
representative GPSR@taylorandfrancis.com
Taylor & Francis Verlag GmbH, Kaufingerstraße 24, 80331 München, Germany

www.ingramcontent.com/pod-product-compliance
Lightning Source LLC
Chambersburg PA
CBHW070727220326
41598CB00024BA/3335